FINANCE AND SECURITY

MARTIN S. NAVIAS

Finance and Security

Global Vulnerabilities, Threats and Responses

HURST & COMPANY, LONDON

First published in the United Kingdom in 2019 by
C. Hurst & Co. (Publishers) Ltd.,
41 Great Russell Street, London, WC1B 3PL
© Martin S. Navias, 2019
All rights reserved.
Printed in the United Kindgom by Bell & Bain Ltd, Glasgow

Distributed in the United States, Canada and Latin America by
Oxford University Press, 198 Madison Avenue, New York, NY 10016,
United States of America

The right of Martin S. Navias to be identified as the author of
this publication is asserted by him in accordance with the
Copyright, Designs and Patents Act, 1988.

A Cataloguing-in-Publication data record for this book
is available from the British Library.

ISBN: 9781787381360

This book is printed using paper from registered sustainable
and managed sources.

www.hurstpublishers.com

For Gabriel and Eve

CONTENTS

ABBREVIATIONS

AML	Anti-Money Laundering
AQAP	Al-Qaeda in the Arabian Peninsula
AQIM	Al-Qaeda in the Maghreb
ATCSA	Anti-Terrorism Crime and Security Act 2001
BBA	British Bankers' Association
BIS	Bureau of Industry and Security
CAATSA	Countering America's Adversaries through Sanctions Act 2017
CCL	Commerce Control List
CFT	Counter-Terrorism Financing
CIFG	Counter-ISIL Finance Group
CTC	Counter-Terrorism Committee
CTITF	Counter-Terrorism Implementation Task Force
CTU	Counter-Terrorism Unit
DAML	Defence against Money Laundering Request
DPs	Designated Persons
EAR	Export Administration Regulations
ECCN	Export Control Classification Number
ECJU	Export Control Joint Unit
FARC	The Revolutionary Armed Forces of Colombia
FATF	Financial Action Task Force
FCA	Financial Conduct Authority
FCPA	Foreign Corrupt Practices Act 1977
FIU	Financial Intelligence Unit
FinCEN	Financial Crimes Enforcement Network
FSA	Financial Services Authority
FTF	Foreign Terrorist Fighters

ABBREVIATIONS

GEA	General Export Authorisation
GST	Goods, Services and Technologies
HMRC	Her Majesty's Revenue and Customs
HVD	High Value Dealers
IFC	International Financial Centre
ILSA	Iran–Libya Sanctions Act 1996
IRGC-QF	Iranian Revolutionary Guard Corps—al-Quds Force
ISIS	Islamic State in Iraq and Syria
ITAR	International Traffic in Arms Regulations
ITSR	Iranian Transactions and Sanctions Regulations 2012
JCPOA	Joint Comprehensive Plan of Action
JMLIT	Joint Money Laundering Intelligence Taskforce
KFR	Kidnapping for Ransom
MCLA	Money Laundering Control Act 1986
NCA	National Crime Agency
NECC	National Economic Crime Centre
NGO	Non-Governmental Organisation
NRA 2017	*National Risk Assessment of Money Laundering and Terrorist Financing 2017*
NTFIU	National Terrorist Financial Intelligence Unit
NTFRA	National Terrorist Financing Risk Assessment
OFAC	Office of Foreign Assets Control
OFSI	Office of Financial Sanctions Implementation
OGEL	Open General Export Licence
OIEL	Open Individual Export Licence
PEPs	Politically Exposed Persons
PKK	The Kurdistan Workers' Party
PLI Model	Placement, Layering and Integration Model
POCA	Proceeds of Crime Act 2002
PML	Professional Money Laundering
RBA	Risk-Based Approach
SAMLA	Sanctions and Anti-Money Laundering Act 2018
SAR	Suspicious Activity Report
SDNs	Specially Designated Nationals
SFO	Serious Fraud Office
SIEL	Standard Individual Export Licence
TACT	Terrorism Act 2000
TAFA	Terrorist Asset-Freezing Act 2010

ABBREVIATIONS

TBML	Trade-Based Money Laundering
UNCAC	United Nations Convention against Corruption
UNCCT	United Nations Counter-Terrorism Centre
UNSC	United Nations Security Council
UNSCR	United Nations Security Council Resolution
USML	United States Munitions List
WMD	Weapons of Mass Destruction

INTRODUCTION

The Meaning of Finance and Security

Security can mean very different things to different people. Both finance and defence professionals frequently employ the term but the concepts to which they are referring are essentially distinct.

When bankers, lawyers and others involved in finance discuss 'security' they are often describing the process whereby a creditor—for example a bank—will take a legally recognised stake in the assets of a borrower. Should that borrower fail to perform its contractual obligations to the bank, such as repaying a loan, then the bank has a right to appropriate those assets in order to make good monies owed to it by that borrower. In support of this process, English law, for example, recognises several types of legal 'security'—mortgages, charges, pledges and liens—but the types of legal and commercial security that can be created vary by jurisdiction.[1] Concomitantly, in a financial context, a 'security' may also refer to a tradable financial instrument that has monetary value. Securities here may be classified as equity securities such as shares and debt securities such as bonds and debentures.

In the fields of war studies, international relations and politics, security carries a totally unrelated connotation. It can mean national and personal well-being, concepts that the layman can readily appreciate but which also have a particular sense and construction within the defence and security space. Thus, the security of the nation state is a service usually primarily provided by a government which develops policies, strategies and forces to protect its institutions, its economy and its populace against a range of military, non-military, conventional and non-conventional threats.

These two communities, finance professionals and defence and security professionals, often work in parallel realms with little overlap and rare meet-

1

ing. However, certainly since the 11 September 2001 terrorist attacks on the United States, the international finance community has increasingly been tasked by the state to help defend the financial system against a range of emerging global threats. Similarly, the security community has found itself drawn further into the world of finance as the global financial system has been recognised not only as a critical node of state power and national interest, but also as one of acute vulnerability.

These growing areas of overlap require both financiers and security personnel to broaden their expertise because today's threats are not narrowly financial nor are they solely political or military. Personnel in both fields require an appreciation of financial products, services and processes as well as an understanding of security threats such as terrorism, corruption and proliferation and how these fit within national security and foreign policy. Those working within the relevant departments of state as well as students of security policy also need to grasp some of the essentials of global financial practices and policies, at least to the extent that those relate to today's key security challenges. A more holistic approach that integrates the study of security risks with that of the uses of finance as defensive and offensive security weapons will enhance institutional compliance while also educating defence and security professionals about the difficulties facing those working in finance who are increasingly tasked by the state with combating financial security threats.

The Financial Sector and Its Vulnerabilities

The financial sector is that segment of the economy that consists of a range of markets, instruments and institutions which together facilitate the conduct of financial transactions such as investments, payments and savings. Key financial institutions such as central banks, retail and commercial banks, investment funds, insurance companies, stock brokerages, leasing companies, state-sponsored enterprises, pension funds, credit unions, rating agencies, credit information agencies and many other types of organisations and institutions manage money and provide commercial services to both retail and commercial customers. They play an intermediary role by directing funds towards investment and thereby directly affect the wider economy. The financial markets consist of capital markets such as those dealing with debt instruments and equities as well as money markets where short-term debt instruments are traded. There are numerous instruments available within developed financial sectors and these include loans, debentures, bonds, shares and other forms of specialised debt and equity instruments. All these institutions, instruments and markets are firmly embedded within a legal and regulatory framework.

The proper and fair functioning of the financial system is key to the broader or 'real' economy. When running smoothly, the financial sector contributes to the well-being of the state by enhancing resource utilisation, by enabling risky investments through both pricing and pooling resources, by encouraging savings in financial assets by ensuring relevant and attractive yields, and by reducing costs and thereby helping increase transactions. As noted in one Swedish study:

> The adequacy of financial institutions, instruments and markets can affect the volume of financial savings mobilised and the efficiency with which they are allocated to productive uses. ... It is the task of the government to create this confidence and to enable the environment for the operators in the financial markets.[2]

International financial centres consist of the agglomeration, for political, economic and geographical reasons, of banks and other financial intermediaries. According to one study, for an international financial centre to be recognised as such it must manifest a concentration of all of the following five banking activities: foreign borrowing and lending; foreign investment; transferring funds across political boundaries; currency exchange such as foreign exchange trading; and the financing of foreign trade.[3] London and New York are pre-eminent examples, but others exist in North America, Europe and Asia. All these centres, and the financial sectors functioning within them, play a critical role in respect of each of their respective jurisdictions' economic strength as well as, singly and together, the power and dynamism of the global economy.

Take, for example, the United Kingdom's financial sector. This consists of banking, securities dealing, fund management and derivatives as well as insurance, all supported by legal, accounting and management consulting services. The sector is undoubtedly critical to the UK economy, to which it contributed £174.3 billion in 2016, or 10 per cent of the total (6.6 per cent from financial services and 3.4 per cent from related professional services).

There are over 19,600 financial institutions of various types operating in the United Kingdom with 87.5 per cent of the total market share consolidated into six banks. Approximately 1,800 principal money service businesses are registered with Her Majesty's Revenue and Customs providing services such as currency exchange, money transmission and check cashing. The United Kingdom is a world financial centre for securities issuance with nearly 5,700 retail investment firms. The insurance sector consists of over 650 general insurance firms and over 230 pension and income retirement firms. In addition, there are approximately 18,600 firms providing legal services and 48,000 providing accountancy services. There are more than 300 registered casinos in the country, more than 10,000 registered estate agent businesses and more than 700 so-called high value dealers trading in expensive and luxury consumer goods.[4]

Tax revenue and employment also make significant contributions here. Thus, the United Kingdom boasts the world's fourth-largest banking sector, fourth-largest insurance and second-largest legal services sector. It is a leading global centre for fund management and global wholesale financial markets. Britain controls 37 per cent of the global total of foreign exchange trading and 39 per cent of over-the-counter derivatives trading. It plays a significant role in cross-border bank lending. It is also a major European centre for private banking, hedge funds and private equity. Its trade surplus in financial services is almost the equal of the combined surpluses of the next three leading jurisdictions.[5] The stability and security of such a sector must surely rank high as policy objectives for those concerned with both the nation's security and the global financial system, with which the nation's financial system is tightly integrated.

Of course, the United Kingdom's financial market is dwarfed by that of the United States. The latter has a massive and diverse financial sector, and the importance of that sector within the international economy together with the global dominance of the US dollar results in trillions of dollars moving through the US banking system on a daily basis.

The US banking sector consists of approximately 13,000 depository institutions of which about a half are banks and the other are credit unions. Five banks hold approximately 40 per cent of total domestic deposits—JPMorgan, Wells Fargo, Bank of America, Citigroup and U.S. Bancorp.

Given this enormous economy and its global significance, the US financial system is particularly vulnerable to money-laundering activity stemming from illegal cross-border flows of money and both foreign and domestic predicate offences. As noted in the United States *National Money Laundering Risk Assessment 2015*:

> The size and sophistication of the U.S. financial system accommodates the financial needs of individuals and industries globally. The breadth of products and services offered by U.S. financial institutions and the range of customers served and technology deployed, creates a complex, dynamic environment in which legitimate and illegitimate actors are continuously seeking opportunities.[6]

The financial system whose vulnerabilities we are concerned about here is not simply national but global. And globalisation in its more current manifestation is a relatively recent phenomenon that emerged rapidly from a far more staid, stable and essentially protected financial environment. In the 1960s and 1970s the City of London was a very different place from what it is today. The popular image is of an old-boys' network staffing institutions, long Martini lunches and fixed hours of business may have been somewhat of a caricature, the banking system was extremely tightly controlled and highly regulated such that while large institutions could engage in various international transactions, the capacity to engage in regular cross-border dealings was restricted. Dynamism was subordinated to stability and control. The Margaret Thatcher-inspired 'Big Bang'—the sudden deregulation of financial markets in 1986—changed everything. The major deregulation of the financial sector that ensued led to significant alterations in the way banks functioned and how they accessed capital. Retail banks purchased investment banks and stockbroking arms, and, with increased capital to invest, these so-called 'universal banks' sought opportunities outside their traditional national boundaries and began offering retail and commercial customers a much wider range of financial services in new markets. The effect of Big Bang was to better enable London to challenge its New York competitors, though of course these new developments had already begun to gather steam in the United States as well.

In the wake of the Great Depression, which engulfed the United States during the late 1920s and early 1930s, the US government passed the 1933 Glass–Steagall Act which sought to protect both consumers and the economy by separating retail/commercial banking from investment banking. This Act was ultimately repealed in the late 1990s, though truth be told it had been weakened long before then. The consequences were not dissimilar from what was happening across the Atlantic. Large-scale mergers and acquisitions within the banking sector followed quickly. The emergence and rapid expansion of JPMorgan Chase, Citigroup and Bank of America led to these three institutions having combined assets of approximately \$6.5 trillion or 36 per cent of the US gross domestic product.[7] The range and sophistication of financial instruments on offer multiplied exponentially from simple 'vanilla' loans, to various type of bonds, repos, swaps and other increasingly exotic derivative instruments, to name but a few. Not surprisingly, the ability of regulators to manage this fast-changing financial institutional landscape became tested more than ever before.

Equally as challenging for those institutions and authorities tasked with regulating and protecting the financial sector was the fact that the enormous amounts of energy and capital that were released were not simply the result of deregulation. They were also a consequence of concomitant technological changes that permitted massive improvements in computing power. The emergence of the internet and consequently screen-based trading made it possible for extremely speedy and complex transactions to be carried out with great ease and dexterity across the globe. Increased availability of capital coupled with these technological innovations coincided with mounting international interest in cross-border investments. This process was further driven by significant rises in the pools of savings and similar processes of deregulation in a growing number of international markets. Both US- and European-based money and financial services progressively spread to a range of far-flung jurisdictions, some of whose regulations and controls were decidedly less rigorous than those of New York and London, bringing with them serious limitations and problems.

Attracted by the internationalisation and profits on offer, traditional banks were soon accompanied in the market by newly arrived financial intermediaries such as private equity funds, hedge funds, large supermarket chains, insurance companies, pensions funds and the like, which began to offer a range of services, many in competition with existing financial institutions. By the end of 2001, the market capitalisation of the world's fifteen-largest financial services providers included four non-banks. The regulatory demands to ensure that these non-bank financial institutions met similar standards as traditional banking organisations placed additional strain on the existing legal and regulatory frameworks. The emergence of technologies such as internet banking and, more recently, Blockchain have further added to the vulnerabilities of the sector.

The combination and confluence of all these developments have undoubtedly had enormous advantages for national, regional and global economies. When we refer to globalisation we mean largely that of the financial and related sectors. But the speed of these developments has in several areas outpaced the ability of authorities to monitor and control them, leaving the system vulnerable to criminal, terrorist and other forms of penetration which inevitably seek out the weakest links in order to achieve their goals.

Key Threats to the Financial System

This global financial sector is vulnerable to a range of primarily economic-related threats. The sub-prime mortgage crisis of 2008 and its ultimate impact upon the global economy provide a case in point. What causes these shocks and how they may ripple across the integrated global system is the subject of study by economists. What we as students of war and security studies in general, and of Finance and Security in particular, are concerned with is how malevolent actors such as criminals, terrorists, corrupt individuals and proliferators may exploit the financial system for their ends, thus impugning its integrity and stability and that of the broader political system too. What must also interest us is how we can use financial institutions, markets and products—ultimately the core elements of our financial power—to support and prosecute domestic and foreign policy and military goals.

Brigitte Unger and colleagues have analysed the impact of money laundering on the economy with specific reference to the Netherlands, but their findings are also relevant to understanding the impact on the global financial system of security threats. These can include direct losses to both victims and society of criminal activity, distortion of consumption and savings, distortion of investment with unfair competition and artificial price increases, alterations in both imports and exports, impacts upon national growth, negative impacts upon income, employment and output, decreases in public sector revenues, threats to privatisation, negative reputational effects on the financial sector, distortion of statistics, rise in crime, the undermining of political institutions, the undermining of foreign policies, and increase in terrorism.[8] The impact of such threats is broad and pervasive and by undermining the integrity of the financial system they serve to undermine the security of the state.

That threats to the financial system represent threats to broader national security interests is increasingly being articulated within national security strategies. Thus the Obama administration's 2015 *National Security Strategy* conceives of terrorism, crime and corruption as interconnected and mutually reinforcing threats.[9] That same year, in testimony before the US Senate Committee on Armed Services, Director of National Intelligence James Clapper identified terrorism and transnational organised crime as among the top eight international threats to US national security.[10] Similarly, the 2015 US *National Terrorist Financing Risk Assessment* states categorically that 'the fight against money laundering and terrorist financing is a pillar of United States national security and a strong financial system'[11] and that the purpose of the assessment is to identify and combat relevant threats and to understand the risks posed by such threats to both the US financial system and national security.

In 2017 the Trump administration's articulation of its national security strategy, while differing in tone and substance from that of its predecessor, was even more explicit in seeking to protect the American financial system from enemy attack. Thus, it is stated as part of the *National Security Strategy of the*

United States of America that 'we will disrupt the financial, materiel, and personnel supply chains of terrorist organisations. We will sever their financing and protect the United States and international financial systems from abuse'.[12] As to transnational criminal organisations, the 2017 strategy document notes that these must be dismantled as they may undermine democratic institutions and pose various national security threats including terrorism. In addition, it argues that transnational criminal organisations may in fact be used by state adversaries as instruments of national power.[13] There is also a focus on the specific dangers posed to US businesses by the threat of corruption. Using its diplomatic and economic tools, the United States will therefore continue to target corrupt foreign officials and work with other countries in the fight against corruption so that US businesses can better function in transparent business environments.

Similarly, in the United Kingdom, key national security pronouncements and risk assessments recognise the importance of securing the financial sector against a range of threats and there is also a clear perception of the consequences of failing to do so. The UK's 2015 *National Risk Assessment of Money Laundering and Terrorist Financing* notes that the size and complexity of the UK financial sector makes it more exposed to criminality than financial sectors in many other countries and that terrorist financing poses a significant threat to UK national security.[14] This is reinforced in the 2017 iteration of the assessment, which states that the government recognises that 'the U.K.'s openness and status as a global financial centre exposes it to the risk of illicit financial flows'.[15] Again, like the United States, the United Kingdom views terrorist financing and criminal financial activity as mutually reinforcing threats. The 2015 *National Security Strategy and Strategic Defence and Security Review* identified terrorism and serious and organised crime as 'Tier One' and 'Tier Two' national security threats respectively.[16] In the UK's 2016 *Action Plan for Anti-Money Laundering and Counter-Terrorist Financing* it is explicitly underlined that 'money laundering and terrorist financing undermine the integrity of our financial institutions and markets'.[17] Given the centrality of those institutions and markets to the United Kingdom's economic health and progress, there is an unambiguous articulation of the view that a threat to this sector represents a threat to the country's broader national security interests.

Responses

The study of Finance and Security involves not only the threats to a country's financial system but also the development and co-ordination of responses to those threats. The goal of implementing such responses is to protect the financial system from exploitation and abuse and by so doing enhance the integrity of the financial sector and thus contribute to both safety and security. Because the UK financial sector is global in its reach and interconnectedness, the defensive response has to be similarly global in its structure and ambitions. If it were not so, then criminals, terrorists and other malign actors could readily penetrate the weakest link in any globally interconnected defensive network. How the United Nations, the institutional embodiment of multilateral activity, has sought to combat the problems of criminal money laundering and terrorist financing is thus of undoubted import, but of even more significance is the bespoke inter-governmental structure, the Financial Action Task Force (FATF), which was established to recommend and monitor standards of anti-money laundering and counter-terrorist financing. The financial sector essentially resides within nation states and is critical not only to global financial stability but to national security as well, so while global responses may set global and national guidelines and objectives, it is at the level of the nation state that those guidelines are given legal and structural expression and are ultimately enforced.

International guidelines related to anti-money laundering (AML), counter-terrorist financing (CFT) and other policies responding to threats such as financial corruption and proliferation finance are thus interpreted through a national prism, and hence in many respects organisations and processes may differ considerably between jurisdictions. This study will focus primarily on the approach of the United Kingdom but will also refer to key policies of the United States. The reason for this choice is threefold.

First, these two countries' financial sectors are obviously, on several key indices, the largest (or in the case of the United Kingdom, among the largest) and most international and developed in the world. More particularly, so too are their AML and CFT laws and structures. It is also true that an understanding of the possibilities of financial warfare and the defensive and offensive employment of financial weaponry receives its most explicit expression in these two jurisdictions.

Secondly, the United Kingdom's AML and CFT, sanctions and counter-proliferation finance approaches, for a myriad of organisational and policy reasons, not least because of the historical significance of EU regulations, bear similarities to those of other European states and will do for some time to come, Brexit notwithstanding. In this regard the UK model may serve as an imperfect but still relevant and useful template for comprehending the AML and CFT approaches of other EU states, and even countries further afield.

Finally, all students of Finance and Security should have an appreciation of the United States' AML and CFT, sanctions and arms embargoes architecture, not only because of the size of the US financial sector but also because American law in this realm has critical extraterritorial effect. While we will further investigate this subject in Chapters 5 and 6, suffice it here to point out that US policy and enforcement can sometimes reach into and impact upon jurisdictions beyond the United States and directly affect persons who are not American citizens.

The Emergence of Financial Warfare as a Strategy

AML and CFT mechanisms that form the key of Financial Action Task Force recommendations and guidelines are essentially defensive financial weapons. They are put in place by governmental authorities and public and private institutions to defend financial institutions, the financial system and broader national security against potential depredations by criminals, terrorists and proliferators who may seek to exploit the financial system to hide and move their funds for either criminal or terrorist ends. These defensive financial weapons are responsive to changing threats and their existence is based on the presumption that there are actors constantly attempting to gain access to and manipulate the financial sector, be that at the national or global level.

But finance can also be turned into an instrument designed to protect the system and to more forcefully and aggressively prosecute foreign policy and security goals. Indeed, that finance can be employed as a strategic weapon and that financial warfare can amount to a strategy is best demonstrated not only in its defensive but also in its offensive mode.

The most explicit articulation of this approach can be found in Juan Zarate's book *Treasury's War: The Unleashing of a New Era of Financial Warfare*.[18] Zarate served at one point as US Assistant Secretary of the Treasury for Terrorist Financing and Financial Crimes. He tells how the US Treasury sought to combat terrorist financing, develop anti-money laundering systems and expand the power of the Treasury to support national security interests. He is emphatic about how financial power can be used to wage war against America's enemies:

> The use of financial power and influence has become an accepted and central tool for protecting and projecting the national security interests of the United States. Just about every day, the U.S. government employs some form of financial pressure—sanctions, financial and diplomatic suasion, regulatory pressure, or

prosecution—to address issues in every corner of the globe—from terrorism in the Middle East to the drug wars in Mexico or human smuggling in Asia.[19]

Zarate eschews the terms 'defensive' and 'offensive' use of power when he speaks of financial warfare, though he does refer to 'protecting and projecting the national interest', and when he writes particularly about the movement from 'financial to strategic suasion' his emphasis does appear to be on the offensive side.

It is important to recognise that financial warfare, the 'strategic suasion' to which Zarate refers, is construed here in a particular sense, involving more than the traditional employment of sanctions, which has long had a role in foreign policy. At its core it involves the building by the state of a coalition with private enterprise. As Zarate notes: 'What made this type of financial warfare different from and more effective than traditional sanctions was that we harnessed the private sector's own interests and calculus to isolate rogue financial actors.' He goes on to emphasise: 'The banking community was not an ancillary part of American power, but a seminal and central part of it'.[20] This public–private coalition is central to both offensive sanctions and embargo strategies and defensive AML and CFT strategies.

In this sense financial warfare is indeed a strategy because, as Freedman has noted, a core element of strategy involves coalition-building. In his description of what constitutes a strategy, he writes that 'the most effective strategies do not depend solely on violence—but benefit instead from the ability to forge coalitions'.[21] So in this sense Zarate is entirely correct to refer to the use of financial instruments as a strategy, and when viewed in terms of the explicit coalition-building exercise involving state and private enterprise, he is also correct to refer to financial warfare as a new strategy as well.

It is the creation, maintenance and direction of this public–private coalition that represents one of the main themes of this book.

What constitutes the make-up of this AML–CFT financial public–private coalition? Public sector stakeholders here include the various intelligence and law enforcement authorities, dedicated financial intelligence units, financial and also non-financial supervisory authorities, licensing authorities, customs and export and trade control agencies, and other government departments and agencies which have relevant AML–CFT as well as sanctions and related functions. Within the private sector important stakeholders can include banks in their many forms, but also money value transfer institutions, insurance companies, company formation and service providers, trading organisations, non-profit organisations, and professionals such as lawyers and accountants.

Once this public–private financial coalition is formed and functioning, then finance warfare may be more properly waged to protect the system through AML–CFT mechanisms but also more ambitiously and offensively through employing financial sanctions and embargoes. Especially in this offensive context (but, as we shall note, certainly not limited to it), coalition-building involves not only public–private partnerships but also traditional international coalition-building among both state and private institutions, this time in the field of finance.

In recent years these offensive financial weapons of warfare have become increasingly popular, especially for the United States and its allies, as one set of instruments to be employed against their enemies. Students of Finance and Security need therefore to understand the mechanisms of financial sanctions as well as that of embargoes and export controls in their more recent form.

The building of the coalition between the state agencies and private sector financial institutions, non-financial institutions, professionals and others is the first stage of both the offensive and defensive expressions of this new form of warfare. But it is telling that beyond the declaratory policy of all those involved, it has not always been a coalition of the willing, either in its generation or maintenance. Indeed, in the context of this new mode of warfare, the first targets of state financial suasion were not criminals, terrorists or proliferators, but rather banks, other financial and non-financial institutions, and various professionals such as lawyers, accountants and estate agents.

Zarate does not express this reality in such bracing terms—he is far too diplomatic—but both a reading of his study and experience in the field leads to the ineluctable conclusion that the coalition is kept in line not only with calls to patriotism and duty but with varied threats, both explicit and implicit, to self-interest.

This is not in any way to impugn the patriotism or civic-mindedness of bankers and other professionals in committing themselves and their institutions to the global war on crime and terror. However, their calculus is inevitably to balance the requirements of these campaigns with the exigencies of doing business.

The key driver here in public–private coalition-building is that private institutions can in Zarate's words be 'leveraged' when their interests align with the state. In leveraging that alignment the state threatens reputational damage, significant financial penalties and incarceration as weapons in incentivising and coercing coalition-building. 'Financial legitimacy and reputation became the coins of the realm,' he writes, but it is the state that ensures that legitimacy and reputations are unambiguously impugned when potential coalition partners step out of line.

Why financial institutions would baulk at signing up as ready participants in the financial war stems from several pressing factors. The British Bankers' Association for example has estimated that the cost to their members of financial compliance is about £5 billion per annum on core financial crime compliance,[22] including 'enhanced systems and controls and recruitment of staff.'[23] The United Kingdom Financial Conduct Authority reports in a study published in late 2018 that the financial services industry is spending in excess of £650 million a year in dedicated staff time dealing with money laundering, fraud and various financial crimes.[24] In addition, there are problems posed for client relations and business in implementing the rigorous checks and procedures required by the state. There is also significant scepticism about the impact of some of these very expensive and cumbersome efforts in achieving real objectives in the war against money launderers and especially against terrorists and proliferators.

Nevertheless, any doubts these front-line private institutions may have about the overall effectiveness of the campaign are more than swamped by the public relations difficulties posed by state-conducted naming and shaming campaigns, which have led to reputational damage, loss of share value, and significant remediation costs.

Equally as pressing have been the increasingly large fines levied on financial institutions that have infringed AML, CFT and sanctions policies. Over the past few years banks have been fined in excess of $350 billion for breaches of money laundering, terrorist financing, and sanctions breaches. In 2012, the US authorities levied a fine of approximately $1.9 billion on HSBC in connection with the bank's failings in respect of money laundering of Mexican drug money and sanctions violations in relation to Iran, Libya, Sudan, Myanmar and Cuba. That same year, fines were imposed on Standard Chartered ($667 million) for breaches of Iranian sanctions, on ING ($619 million) for breaches of Iranian and Cuban sanctions, and on Credit Suisse ($556 million) for breaches of Iranian sanctions. In 2014 BNP Paribas was fined $8.9 billion for breaches of Iranian and Sudanese sanctions while in 2015 Deutsche Bank was fined $258 million for breaches of Iranian and Syrian sanctions.

The aggressive response of authorities, especially American authorities, to coalition breaches and, most significantly, European banking breaches—no doubt *pour encourager les autres*—have kept most international banks in line with shifting national and global sanctions and arms embargo policies and focused their minds on money laundering, terrorist financing and proliferation financing threats.

Pressures of this kind emanate not only from US authorities but also those of the United Kingdom where the focus tends to be primarily on anti-money laundering failures. For example, in 2013 the UK's Financial Conduct Authority, the key regulator of financial services, levied fines of over £4 million on EFG Private Bank for 'failing to take reasonable care and maintain effective AML controls for high-risk customers'; in 2014 it fined Standard Bank plc (South Africa) £7.6 million for failings related to AML policies and procedures for customers connected to so-called politically exposed persons; in 2015 it imposed fines of £72 million on Barclays Bank for failing to carry out proper AML and corruption checks on billions' worth of transactions carried out by politically exposed persons; and in 2017 it imposed £163 million in fines on Deutsche Bank AG for inadequate AML controls.[25]

It should thus be in the interests of those working in the private sector to have some overview of the key political, legal and financial factors that drive financial warfare strategy at the national and multilateral levels. Moreover, it may be useful for those active at the state policy level to gain greater awareness of the difficulties and costs encountered by those tasked with conducting policy in front-line private institutions.

In Chapter 1 we will analyse the key principles of money laundering and demonstrate the threat it poses to financial security. In Chapter 2 we will focus on the security challenges posed by terrorist financing and the differences between this model and that of criminal money laundering. Chapter 3 will review how both criminals and terrorists move funds across the international financial system including their exploitation of both the informal and formal financial sectors. Chapters 4 and 5 will review global and national responses to these threats to the security of the financial system. In Chapter 6 we change gears and look at how financial weapons such as financial sanctions and arms embargoes can be offensively used to further foreign policy and security goals. Chapter 7 reviews the particular problem posed by proliferation finance and the steps taken to control its manifestation. The conclusion will attempt to assess the overall efficacy of these defensive and offensive financial mechanisms in enhancing financial security.

The problems of criminal financing have long been addressed in the academic criminology and development studies spaces. As sanctions and embargoes gain growing import as foreign policy tools, studies of their efficacy are increasing within the field of international relations. This study attempts something different. It is based on an MA course I teach in the Department of War Studies, King's College London, and reflects my interests in war studies and international relations and some of my professional experiences as a banking and finance lawyer in the City of London. The aim is to try and integrate the separate areas of money laundering, terrorist and proliferation financing, sanctions and export controls into a coherent whole as viewed through a war studies and practical compliance prism. It seeks to provide a text for both practitioners and students of the emerging field of Finance and Security while at the same time legitimising the treatment of this subject as a distinct area of focus within the broader realm of the study of security and war.

1

THE LAUNDERING OF MONEY

Money laundering is the process by which criminally derived funds are transformed into part of the legitimate economy.

E. — *Breaking the link between the crime and the "good life"*

The Black Market Peso Exchange Scheme: A Nexus of Criminality and Commerce

Hollywood and, more recently, streaming services such as HBO and Netflix continue to cater to our fascination with the global drugs trade. The extreme violence, the lavish lifestyles of the top dealers, and the frightening criminality and hard-edged police work to be found at the sharp end of the business have provided grist for a host of movies and television series from *Scarface* (1983) and *Blow* (2001) to *Breaking Bad* (2008–13), *Narcos* (2015–17), *El Chapo* (2017–) and the BBC's *McMafia* (2018–) to name but a few. As these demonstrate to varying degrees, the movement of marijuana, cocaine and other drugs from the centres of manufacture to eager and receptive markets often located across hostile borders gives rise to unimaginable riches. This in turn precipitates the inevitable requirement that these funds be somehow integrated into the legitimate global economy and rendered useable by those seeking to profit from this illegal business.

The insatiable North American demand for drugs, which took off in the 1960s and expanded dramatically during the following decades, meant that the South American drug lords were soon generating sales in the billions of dollars.[1] The profit margins on drug sales were also truly massive. Writing about

Pablo Escobar, his accountant Roberto Escobar noted the extreme profligacy that emanated from the huge tidal wave of cash that flowed from the drug sales: 'Pablo was earning so much that each year we would write off 10 per cent of the money because the rats would eat it in storage or it would be damaged by water or lost.'[2] The drugs poured into the United States by way of aircraft, ships, trucks and people but the frustration was that while massive profits were being generated as a result of street distribution, the vast amounts of cash that were being garnered daily were not immediately accessible to the producers.

The drug lords in Colombia rarely ventured from their secure bases, be that Medellin in the case of Pablo Escobar or Cali in the case of the Ochoa brothers. While they had expenses in the United States, costs that were not insignificant due to the demands of distribution, security and bribery, their real needs were closer to home in Colombia—these needs again being the local costs of distribution, security and bribery, but ultimately of course also expenses associated with the good life. There was little to be gained given all the risks involved and efforts undertaken if most of the profits remained near the nodes of sale in urban North America rather than in the marketplaces of Colombia.

Another challenge facing the Colombian drug lords was that while the drug trade profits were being generated in American dollars, these dollars, even if successfully repatriated to Colombia, were of limited use because local business was conducted in pesos. While dollars could have been used to buy and sell things in Colombia, that would not have solved the drug lords' problem, given the huge amounts of cash in dollars that were being generated. These dollar profits mostly had to be converted into pesos, a not insignificant task. Both the movement across borders of enormous quantities of dollars and their conversion into local currency had to take place well below the law enforcement and regulatory radars of both the US and local authorities.

Had the funds themselves been legal, the most obvious method of repatriation would have been simply introducing them into the US banking system and wiring the money home to Colombia. The provenance and huge quantity of cash involved rendered such options impractical as it would have immediately attracted the attention of the authorities, who would seek to 'follow the money' back to the cartels. In the 1970s, as both the price and volume of sales of drugs skyrocketed, the US authorities were far from oblivious to the threat and had begun cracking down on suspicious deposits by requiring banks and businesses to report suspicious transactions on bank deposits in excess of a *de minimis* amount. The US Bank Secrecy Act of 1970, which sought to curtail the vulnerability of the banking system to drug money, was initially only sporadically enforced and banking compliance levels were not always of the highest standard.[3]

An alternative route, the placing of drug sale profits beyond the United States into offshore accounts, also became more difficult in the 1970s and 1980s. So great were the sums involved here that even physical means of transport—the planes, boats and trucks which had brought the drugs into the United States in the first place—proved insufficient for the task of repatriating the funds. The fleets of planes and boats needed to shift the bundles of cash, often in small denominations because that was how street-level drug sales were transacted, to these offshore centres, would have become increasingly visible and vulnerable to interdiction. Authorities in Aruba, the British Virgin Islands, Bahamas and the Cayman Islands, which were often favoured destinations of the cartels for temporarily storing the dollars before moving them back to the United States or to other destinations, also began to become suspicious of this scam. They were unable in any event to ignore the mass influx of dollars into their territories, hence their airport and seaport controls began to be strengthened along with their banking regulations.

To solve this conundrum, the cartels began to rely on a scheme that disguised the illegal origin of the funds while at the same time helping to move the money back to Colombia. This method is a form of trade-based money laundering (TBML) and the so-called Black Market Peso Exchange Scheme, which involved the laundering of these dollar profits out of the United States and into various jurisdictions including Colombia, became the most successful example of this type of money laundering scheme. At its height, the scheme is believed to have moved over $4 billion a year of drug money out of the United States and into the drug-producing states.[4]

It is important to recognise that the drug cartels were not acting alone in these laundering efforts. In Colombia, legitimate businessmen and ordinary citizens became ready accomplices of the drug exporters' trade-based laundering scheme by participating in an arrangement that helped disguise the underlying regional drug transactions. The reasons for the readiness of these legitimate businesses to work hand in hand with drug cartels goes back to the 1960s when the Colombian authorities, seeking to protect the value of the peso made vulnerable by the reliance of the local economy on the coffee crop, restricted Colombian citizens' foreign currency holdings. Should for example a Medellin trader wish to import air conditioners from Miami, the costs could be prohibitive as the Colombian central bank would impose surtaxes on the peso–dollar exchange and sales taxes on the imported cooling systems. Once value-added taxes and customs duties were added, the costs could make the transaction increasingly uneconomical or certainly more unattractive.[5]

This was not a problem restricted to Colombian importers. The official and highly unfavourable peso–dollar exchange and the significant exchange fees affected everyone who wished to visit, holiday or study outside the country. A Bogota doctor who wanted to send his daughter to study at the University of Chicago for example would find himself priced out of the market by the official costs associated with the currency exchange. It was in response to this untapped demand that money launderers working for the drug cartels stepped in. With massive amounts of dollars sitting unusable within the United States, the launderers could offer the Colombian businessman seeking to import air conditioners or a doctor in Bogota seeking to pay his daughter's university fees a far more attractive peso–dollar exchange rate than that offered by the Colombian central bank.

In this sense, the Black Market Peso Exchange Scheme differs from most unofficial black market currency schemes in that the seller of the currency, rather than charging the purchaser a premium (reflecting the additional risk of the illegal transaction), actually charges an amount less than the official rate. Such a situation can only arise because the drug cartel is so cash-rich that it can afford to take a significant cut in its margins in order to gain access to its funds.

There are numerous ways this exchange could work but let us break the simplified process down into five stages, as follows:

- Stage One: The sale of drugs on the streets of US cities nets the Colombian drug lords millions of dollars in small (and bulky) denominations. These funds remain in the physical possession of the drug dealers' local representatives in a myriad of American cities but all far from Colombia, and also in US dollars rather than in Colombian pesos.

- Stage Two: A peso broker working in Bogota wants $1 million for his own exchange transactions in the United States. If he went to the Colombian central bank to purchase the foreign currency, it would cost him the equivalent of say $1 million in pesos (assuming a 1:1 exchange rate) and he would still have to pay fees and taxes on the transaction. Instead he contacts the drug cartel's money-laundering representatives, who agree to physically deliver to the peso broker's agent somewhere in the United States $1 million in exchange for the discounted equivalent of (let us say) $700,000 in pesos delivered by the peso broker to the cartel somewhere in Colombia.

- Stage Three: The drug cartel is now free to use pesos equivalent to $700,000 in Colombia, pesos whose origins are actually dollar drug deals carried out in the streets of North American cities. It has cost the cartel effectively $300,000 but the money is back home in Colombia and in more locally usable pesos.

- Stage Four: The peso broker now assumes the risk of placing the dollars he has acquired from the cartel into his US bank accounts. He contacts the Bogota businessman who wants to import air conditioners from a Miami-based company. The air conditioners cost $1 million and the peso broker is prepared to sell the businessmen $1 million dollars for the equivalent of $800,000 in pesos. The peso broker makes a profit of $100,000 in pesos. The businessman is effectively saving $200,000 by dealing on the black market rather than going to the central bank.
- Stage 5. The peso broker's representative in the United States pays the Miami-based company the $1 million sale price on behalf of the Bogota businessman and purchases the air conditioners. The Bogota businessman imports the air conditioners into Colombia.

This international trade scheme appears legitimate on the surface and provides a ready rationale to explain the appearance of money in the accounts of

BLACK MARKET PESO EXCHANGE SCHEME
A Colimbian Example

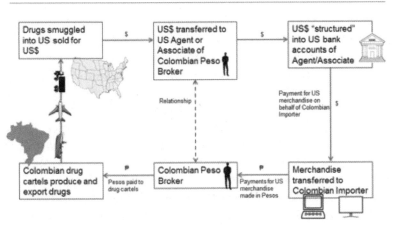

This diagram is based upon one published in *The National Money Laundering Risk Assessment, 2015*, Department of the Treasury, p. 30 and various Financial Action Task Force diagrams.

the drug lords and the peso broker. The Colombian importer similarly has an apparently legal explanation for the source of his funds (the domestic sale of air conditioners he has imported from the United States) while the Miami-based exporter may or may not be aware of the related criminality but similarly appears to have engaged in a legal activity. Any links to the underlying crimes of drug sales and tax evasion (to name but a few of the illegal activities potentially involved here) are hidden under the cover of a legal import and export business.

There are of course multiple variants of this TBML process, but all essentially work by employing trade transactions such as the import and export of merchandise, together with the falsification of the value of the goods, to disguise and cover an underlying criminal transaction. For example, the peso broker may have been a separate party from US-based criminal groups who acted as local American money brokers and dealt with the US side of the business, taking physical control of the dollars from the cartel, placing the monies into bank accounts, and then paying the Colombian-based peso broker a fee as well as settling accounts with the US-based exporter. Indeed, there may be several separate brokers working on different stages of the transaction. Each individual transaction may have been initiated independently and the size of the 'haircut' that each selling party is prepared to take varies depending upon specific deal and general market considerations. Drug monies could also be intermingled with legal receipts of legitimate businesses to further disguise their origins.

The model described here is not limited to Colombia. US and Colombian law enforcement attacks on the Colombian cartels, coupled with various intra-industry changes, saw the locus of drug export activities shift northwards to Mexico.[6] Mexican drug cartels were soon earning billions of dollars from sales into the United States and the laundering of these profits soon became a major focus of the Mexican drug-trafficking organisations.[7]

Between 2007 and 2013, however, the US authorities aggressively sought to sever customer relationships between their local banks and Mexican money exchanges, and these restrictions—together with efforts by the Mexican authorities beginning in 2010 to restrict dollar deposits—complicated efforts to shift monies directly back into Mexico and led to increasing quantities of profits being used within the United States and in third countries.[8] In any event, because of geographical contiguity with the United States, bulk cash smuggling remained a major method of moving profits across the border, something not readily available to the Colombians.[9] Cash not used in the

United States or stored in Mexico (often outside the financial system) was transferred to countries such Panama, Colombia, Brazil, Honduras, Guatemala and Argentina.[10]

Trade-based money-laundering schemes still play a role in shifting profits southwards out of the United States. Thus there have emerged separate Mexican money brokers who trade with Mexican importers in a similar manner to the role traditionally played by Colombian brokers in the Colombian market.[11] The scheme is now also sometimes used by the Mexican cartels to satisfy debts owed by them to any of their Colombian cocaine suppliers who may be forwarding product to Mexico for onward transport to the United States. They do this by delivering cash from their profits in the United States to Colombian agents in the country, who then place the cash into the US financial system, inter alia, to pay American businesses shipping goods to Colombia. Businessmen in Colombia pay pesos to brokers in Colombia, who then use this money to pay the Mexican drug dealers' debts to their Colombian cocaine suppliers. According to US Treasury analysis: 'This connection between Mexican [drug-trafficking organisations] and Colombian money brokers has reduced the cost and risk to both Mexican and Colombian [drug-trafficking organisations].'[12]

By exploiting this TBML technique the Black Market Peso Exchange Scheme expanded into a major money-laundering phenomenon. Consequently, according to Raymond Kelley, a former commissioner of the US Customs Service: 'The Black Market Peso Exchange is perhaps the largest, most insidious money laundering system in the Western Hemisphere. It's the ultimate nexus between crime and commerce, using global trade to mask global money laundering.'[13]

Global Money Laundering: The World's Largest Industry

While the Black Peso Market Exchange scheme has historically been among the most potent money-laundering schemes in the world, money laundering is a far larger industry than that created by the South and Central American drug cartels. It is undoubtedly a major global business.

Money laundering is the inevitable consequence of crimes aimed at profit generation in which the criminal needs to distance himself from the original transgression. It is the process by which criminally derived funds are injected into the legitimate economy so that they appear legitimate themselves. By appearing legitimate, the link between the original criminal activity and the funds is broken, and the criminal enhances his security.

Imagine a colleague robs a bank in a downtown mall and makes off with $5 million. Yesterday he drove a battered Ford but today he arrives at work to tender his resignation in a new Mercedes CLS. The cheerful performance suit he previously wore every day has been replaced with an expensive Armani. A forwarding address of Boca Raton, Marbella or Pattaya is sent to the human resources department.

Questions will inevitably arise as to the origins of his good fortune. The newspapers are full of reports of a major heist in the neighbourhood but the colleague has the documents to prove that the money derives from an offshore account in the Bahamas where long-held investments in New York real estate deals and Bitcoin speculation have realised themselves. The wealth—at least as far as his documents demonstrate—is all above board. The link between the funds and the crime is broken and is replaced with a patina of legitimacy. Any police investigator seeking to 'follow the money' back to the crime will run up against a paper trail including banking records and recent purchases all attesting to the legitimacy of the funds and reflecting a total disconnection from any nefarious activity.

Money laundering is often a result of large profit-generating crimes in which the criminal has to break the link to his original illegality. It tends to be related to the more serious end of the criminal spectrum because if the crime only generates a small amount of monies there will be little need to engage in the difficult and expensive process of disguising the origins of those funds. Small amounts can often be readily used without raising too much suspicion. It is in the areas of tax evasion, fraud (including identity theft, mortgage fraud, retail and consumer fraud, securities fraud and health-care fraud), drugs, corruption, human trafficking, organised crime and public corruption that the requirement to launder money comes mainly into play.

Estimates of the size of this area are notoriously difficult but a United Nations Office on Drugs and Crime (UNODC) report presents some truly staggering figures: in 2009 the sums laundered were estimated to be 2.7 per cent of global GDP, or US$1.6 trillion.[14] Elsewhere UNODC claims that the proceeds from financial crimes in the United States in 2010—excluding massive amounts linked to tax evasion—was about 2 per cent of GDP or about $300 billion.[15] In more recent figures the organisation has suggested the annual figure to be between 2—5 per cent of global GDP, or $800 billion—$2 trillion in current US dollars.[16] In the United Kingdom, the National Crime Agency (NCA) has estimated that between £36 billion and £90 billion may be laundered through the United Kingdom each year, a figure which does not include British banks' international subsidiaries, themselves also potentially exposed to money-laundering risks.[17] Robert Barrington, the executive director of Transparency International UK, told the House of Commons Home Affairs Committee in 2016 that he estimated money laundering via the UK financial system amounted to at least £100 billion per annum.[18]

Whether these estimates represent a true reflection of the scale of the problem is impossible to say, but there can be no doubt that money laundering forms one of the largest industries in the world. A major and incredibly sophisticated global industry has arisen to help break the link between funds and crimes and render the illicit monies apparently legitimate.

The origin of this modern money-laundering industry remains in some dispute. One theory is that the term itself originated in the 1920s and 1930s in Chicago during Prohibition. Bootleg alcohol, gambling and other illegal activities were at the time generating large amounts of funds for criminals such as Al Capone. Like the more modern drug cartels, these groups too had to cut the link between their funds and the original crimes that had generated them, and they did this by setting up legal businesses through which their illegal monies were streamed. Capone established a chain of laundromats which, while ostensibly dedicated to washing clothes, also served as a means to 'wash' or 'launder' his ill-gotten gains. Laundromats were a cash business with the customer paying cash to use the machines. There was no reliable method of calculating how many times the machines were actually used, so Capone's accountants could claim a far greater customer base than was in reality the case. Illegal funds were mixed with actual earnings from the running of the laundromat business and rendered indistinguishable from them.

When questioned by the authorities as to the source of his wealth, Capone could point to an apparently extremely successful laundromat business with each store generating thousands of dollars. Ultimately Capone was convicted for tax evasion rather than for bootlegging, and this may explain the theory that 'money laundering' stems from the actual laundering operation that he conducted to disguise the link between his profits and his crimes.

Whatever the case, the money-laundering business model received a significant boost when Meyer Lansky, a senior US-based Mafia accountant, began diverting sums generated by criminality in the United States to a discrete set of bank accounts in Switzerland. By doing so Lansky exploited the traditional secrecy of the Swiss banking system, which refused to co-operate with US law enforcement or regulatory authorities to reveal the origins (or even the existence) of funds within accounts or the identity of the account holders. Swiss-based money was of limited use to US-based Mafia groups just as US-based money was to Medellin-based drug lords. Consequently, Lansky devised a scheme whereby some of the money was repatriated to the United States in the form of loans, which meant that 'hitherto illegal money could now be disguised by "loans" provided by compliant foreign banks, which could be declared to the "Revenue" if necessary, and a tax-deduction obtained into the bargain'.[19]

Reference to the term 'money laundering' then appeared during the Watergate scandal when President Nixon's Committee to Re-elect the President sent illegal campaign finance contributions to Mexico and then repatriated the money via a Miami company back into the United States. It was the *Guardian* newspaper that reportedly first used the term in this context when it referred to the process as 'laundering'.[20]

Typologies of Money Laundering

The United Kingdom's *National Risk Assessment of Money Laundering and Terrorist Financing 2017* (NRA 2017) distinguishes between several money-laundering methodologies prevalent within the UK—methodologies which aim to break the link between the criminal origin of funds and their desired ultimate apparently legitimate form. These include:

(i)   Cash-based money laundering, which involves the use of cash-intensive businesses to disguise the criminal origins of funds. These are then combined with the use of legitimate services such as money transmission services and retail banking to move the monies. Al Capone's laundromats are a case in point, but other cash-intensive businesses can readily serve a similar function. This methodology of course reflects the prevalence of cash in key predicate crimes such as drug trafficking, where proceeds often start and remain in cash (fraud proceeds, while rarely starting in cash, may be transformed into cash during the laundering process). The key underlying point here is the essential fungibility and anonymity of cash: when legitimate and illegitimate funds are mixed, they are essentially indistinguishable.

(ii) So-called 'high-end money laundering', which involves significant amounts of criminally originated money-laundering funds often derived from overseas corruption and involving tax evasion, major frauds and corruption. Here there is a reliance on UK-based professional services such as lawyers and accountants. The funds are moved through offshore centres and complex financial structures.

(iii) The TBML type, which was described above in the context of the Black Market Peso Exchange Scheme and which exploits the import-export system to disguise and move criminally generated funds.[21]

The *NRA 2017* identifies cash-based money laundering and high-end money laundering as the greatest threats to the United Kingdom but the report admits that in practice it is sometimes difficult to distinguish between the methodologies, especially when they become integrated and mixed up.

In the United Kingdom, for example, one of the most common variants of TBML impacting upon the country's banks involves the abuse of the open account third-party payments system whereby sellers provide credit services to buyers and then transport the goods to the buyers before any payment for the goods is made. The account is then settled by third parties who pay for the goods. This approach is often favoured in parts of the Middle East and Eastern Europe and is in most cases a totally benign payment mechanism. However, as a system it is vulnerable to money laundering as the payment arrangements can be exploited to disguise the identity of the actual payer as well as the actual source of the monies. The third-party payer may be legitimate but sometimes they may seek to pay the account using illegally obtained funds. According to one professional working in the field: 'Third party payment arrangements strike at the heart of [anti-money laundering and other] controls because they undermine the achievement of the necessary levels of customer due diligence by the business, on which the bank may be relying.'[22] If the third-party payment arrangements are essentially undisclosed, a bank cannot be certain that legal and regulatory obligations are being covered. Clearly these kinds of payment arrangements not just present the risk of exposure to potential money laundering but also reflect vulnerabilities to other forms of financial threat as well. Implementing economic sanctions and arms embargoes may be undermined if payments are derived from a non-sanctioned country when in fact the end user is a sanctioned or embargoed entity.

By explicitly distributing illegal goods, fabricating the actual value of the items being distributed, and lying about the purpose of the underlying financial transaction, such activities have an extremely negative impact on legal business operations with which they directly compete. This is not least because merchandise and currencies trade substantially below market rates, making those legal avenues relatively unattractive.

According to the US Immigration and Customs Enforcement (ICE), Homeland Security Investigations (HSI), the largest investigative arm of the Department of Homeland Security (DHS), TBML-imported merchandise—

estimated to be worth billions of dollars annually—may be dumped onto the market at a discount so as to expedite the money-laundering process. Such losses are the costs of doing business for the money launderer but at the same time this process potentially puts at a severe disadvantage legitimate businesses that cannot compete with these discounts. Such actions may further serve as a barrier to entrepreneurship while at the same time diminishing the government's revenue as import duties and other taxes may be avoided.[23] Thus the impact upon wider economic activity can be profound.

The ICE's HSI has established the Trade Transparency Unit to identify global TBML trends, and it has crafted and published a set of so-called 'red flags' which may indicate the existence of a TBML scheme. These may include:

- payments to vendors made in cash by unrelated third parties;
- payments to vendors made via wire transfers from unrelated third parties;
- payments to vendors made via cheques, bank drafts or postal money orders from unrelated third parties;
- false reporting, such as commodity misclassification, commodity over-valuation or under-valuation;
- carousel transactions (the repeated importation and exportation of the same high-value commodity);
- commodities being traded that do not match the business involved;
- unusual shipping routes or trans-shipment points;
- packaging inconsistent with commodity or shipping method; and
- double invoicing.[24]

As with all red flags—not simply those indicating TBML—the existence of one or more of these indicators will not necessarily be conclusive evidence of the existence of money laundering in general or a trade-based scheme in particular. Nevertheless, at the very least, they may be strong pointers to the necessity for further due diligence.

The PLI Model

Whatever the typology, and whether we are referring to the laundering programmes of the 1920s Chicago Mafia, the 1980s Medellin cartel or, more recently, the many Mexican drug organisations involved in money laundering, the general principles underlying money-laundering processes are the same. In fact, most large-scale organisations engaged in laundering operations seeking to break the link between the origins of their funds and their present finances will engage in a three-stage process consisting of placement, layering and integration—the so-called Placement, Layering and Integration Model (PLI Model).

The three phases of the PLI Model are:

1. Placement—in which illicit proceeds are introduced into the financial system in a way that disguises the true ownership of those funds and seeks to avoid activating the financial industries' anti-money laundering systems.
2. Layering—in which there is an attempt to transfer the proceeds of crime through a series of complex financial transactions that serve to disguise the true origins of the funds.
3. Integration—in which the illicit funds are made to re-enter the economy in an altered character and disguised as legitimate assets.

Table 1.1: The Three Stages of Money Laundering: The PLI Model

Placement	Layering	Integration
Goal	Goal	Goal
To insert criminal proceeds into the legitimate financial system	To conceal the criminal origin of proceeds through disguising and distancing by way of complex financial transactions	Integration of the funds into the legitimate economy
Common Methods	Common Methods	Common Methods
• Subdivided cash deposits into bank branches • Currency conversion at bureaux de change • Cash exchange at casinos • Injecting into cash-intensive businesses	• Global wire transfers • Cash deposits into multiple bank accounts • Splitting and merging various bank accounts • Using various types of legal structuring whose ownership is opaque	• Buy assets (property, luxury goods and similar items) • Create fictitious loans • Use criminal proceeds in transactions with third parties

In practice the phases, like the typologies, are not always clearly distinguishable, and some writers have raised the question whether this classic model is too artificial or entirely relevant to real money-laundering processes, especially at the lower levels where the requirements for concealment are far simpler.[25] Nevertheless, the PLI Model provides a useful starting point for the high-level (large amount) money-laundering process and a means by which it can be analysed.

The initial challenge facing the money launderer in the placement phase is to avoid tripping over one of the first lines of the banks' defensive system, namely the suspicious transaction reporting requirements that are triggered should the deposit exceed a *de minimis* amount. Because the close monitoring of millions of transactions (including deposits) that characterise daily banking is impossible, banks are—in the absence of other intelligence or specific suspicions—usually only required to report to the relevant governmental authorities a specific set of transactions that exceed a defined minimum quantity.

One way of avoiding setting off this alarm system is through the 'structuring' of the deposits, which involves breaking up the illicit cash into smaller amounts and placing them into the banking systems at several entry points. The United States *National Money Laundering Risk Assessment 2015* gives several examples of this technique, including that of an Albuquerque firefighter convicted of drug trafficking and money laundering who structured 37 cash deposits and withdrawals so as to launder drug proceeds; a Las Vegas lawyer sentenced for making or assisting in 15 structured deposits as part of a tax evasion scheme; and a New York police officer who allegedly structured numerous cash deposits of between $1,000 and $7,900 into seven bank accounts aimed at laundering money derived from heroin trafficking.[26]

According to the US Treasury Department, one common method of structuring—especially popular in south-western border regions of the United States—involves using so-called funnel accounts in preparation for funnelling drug proceeds earned there back into Mexico. Technically, a funnel account is a business or an individual account which may be located in one geographic area that receives a myriad of cash inputs, in many instances in amounts below the legally required cash-reporting threshold, and then from which the monies are withdrawn in a different geographic region. These withdrawals are often characterised by the short time elapsing between the deposit of the money and the later withdrawals. By employing this method, the reporting thresholds are not breached and the risk of transporting the cash is avoided or at least minimised.

There are numerous instances of this technique being employed. The US *National Money Laundering Risk Assessment 2015* uses the example of the nine alleged members of the Mexican Gulf Cartel charged in 2013. In this case, from 2008 until 2012, distributors of drugs in Florida paid for the narcotics through the use of structured deposits in various local branches of a national bank while the drug suppliers located in Texas made structured withdrawals and then distributed the cash to couriers who physically carried the money across the Texas border into Mexico.[27] This was a large and complex operation worth millions of dollars.

Characteristic of funnel accounts are the use of nominees (sometimes in the literature referred to as 'smurfs') to deposit the funds into each branch, and the relatively quick turnaround from deposit to withdrawal (which may range from a few minutes to a few days). The accounts are characterised by no other activity aside from the deposit and withdrawal. The challenge for the bank is to distinguish these funnel accounts from legitimate concentration accounts held by businesses which can appear on the surface not too dissimilar to funnel accounts. Of course, the complexity of this operation may be avoided if a complicit merchant agrees to receive an excess reporting requirement sum without making the necessary report.

The entry points into the financial system are of course not only banks. The advantage of banks is that they can provide a range of services—not least of all fund movement between branches and even jurisdictions—useful to the launderer. For smaller amounts of money or where the crime has generated funds in forms that are not that of traditional currency, then the placement methods described above may not be entirely relevant. However, to employ large amounts of money—say, to ultimately buy a significant property or a private aeroplane—the launderer must make wire transfer payments or rely on other types of banking services.

At the lower ends of the crime spectrum where the numbers are smaller, exchanging money at the casino or the bureaux de change may suffice and the layering stage of money laundering may arguably be avoided altogether. So, for example, an individual who has stolen $50,000 through an identity theft scheme may take a portion of that sum and go to a casino where he exchanges his cash for gambling chips. He then goes to the table and gambles and may win or lose some money, though that activity is essentially peripheral to his central interaction with the cashier where he exchanges the chips for cash. In so doing he receives new currency from the cashier not linked to the cash with which he purchased the gambling chips and acquires a built-in narrative as to where these actual funds originated. Similarly, he could go to bureaux de change and exchange his original currency for, say, euros or pesos and then at a later stage change back again to his original currency. Here too the linkage to the original crime appears broken.

In fact any cash business could in theory serve a similar function. Cash, being both fungible and anonymous, can readily be injected into such enterprises and mixed with legitimate funds deriving from that business. However, as noted above, for larger amounts, such activities may be inadequate and banking services and products become necessary for the ultimate proper integration of the funds into the legitimate economy.

Of course, once the untraceable cash is deposited into the banking system it becomes traceable and in theory can be connected to the account holder into whose account the funds have been inserted. Consequently, so as to further distance the origin of the funds from both the initial crime and the depositor, the so-called layering process comes into play, through the generation of various layers of consecutive and coterminous financial transactions. According to one analysis, these complex layers of financial transactions aim specifically to 'disguise the audit trail, source and ownership of funds'.[28] Layering can involve, among other things, wiring funds through numerous accounts, some of which may be in different jurisdictions with different levels of reporting and compliance requirements; converting the deposited cash into several monetary instruments such as bonds and other debt instruments; and employing various legal structures to disguise the source and origin of the funds and add to the general complexity of the trail.

Let us now turn to some of the products and entities that can be used to support the layering process. There are in fact numerous products that can abet layering. Take, for example, debit cards, which certainly have the potential for serving as readily usable layering tools. Launderers can layer their funds by transferring them among numerous debit, credit and other prepaid cards. The cards may have online transfer functions which a launderer can enable if he knows the card number and the password, and an online transfer can be initiated. Purchasing transactions via a 'point of sale purchase' at a merchant and then requiring a refund at another outlet will further layer the process.[29] Debit cards can be used to withdraw the cash at ATMs or resell the purchased goods elsewhere. The merchants may be legitimate or part of the syndicate while the launderer can easily take the debit cards across borders with little danger of interdiction, thereby further layering the process.[30]

In addition to financial instruments, the use of various types of legal entities can also help disguise the origin of the funds and complicate the trail. Here launderers may employ front companies, shell companies, shelf companies, and other forms of corporate structures to support this objective.

Front companies are actual functioning businesses that can combine legal operations and illegitimate earnings. This hides the source, ownership and control of the illegal funds—an example of this are the aforementioned laundromats, where the sources of funds were integrated and appeared undifferentiated in the company's accounts. In the United Kingdom another type of business where money laundering is believed to take place is high-street nail bars, which have proliferated in recent years. Sir Stephen Lander, who previously headed Britain's Serious Organised Crime Agency, argued that these businesses were ideal means of money laundering as there were no formal client lists—customers were walk-ins off the street—and this 'means there could be suitcases out at the back stuffed with cash and the owner could simply turn around and say these were profits from the business. There are no records necessary to be kept and thus it's harder to prove the cash came from crimes.'[31]

A shell company, on the other hand, is registered as a legal entity but, unlike a front company, has no real physical operations. It just exists in form only and is not a functioning business. There may of course be a legitimate reason for this, as it may be acting as a holding company for property or financial assets, but they can also be used to store illegally derived assets and to disguise ownership. Igor Angelini, head of Europol's Financial Intelligence Group, has stated that shell companies 'play an important role in large-scale money laundering activities' and that they are often a means to 'transfer bribe money.'[32]

The investigations by a group of journalists into the so-called Panama Papers certainly supported this latter assertion and revealed the significance of shell companies in the layering of illegitimate funds. The International Consortium of Investigative Journalists, together with the German newspaper *Süddeutsche Zeitung* and many other media organisations, investigated approximately 11.5 million leaked documents which exposed the offshore holdings of politically exposed persons linked to several illegal activities including money laundering. The documents represented one of the biggest leaks of inside information in history and included almost forty years of data from a law firm based in Panama, Mossack Fonseca, a major creator of shell companies.[33] Their investigations revealed the offshore holding of about 140 politicians and offices including more than 214,000 offshore entities connected to people in more than 200 countries. For example, the Panama Papers revealed that corrupt government officials across Africa employed shell companies in several offshore jurisdictions to hide profits from the sale of natural resources and the bribes paid to gain access to them. Owners of the shell companies included three Nigerian oil ministers, as well as senior employees of the national oil company and two former state governors. In addition, owners of diamond mines in Sierra Leone and safari companies in Zimbabwe and

Kenya also set up shell companies with no other function than storing and disguising their illegally derived assets.[34]

These entities should be differentiated from shelf companies, which are simply entities that are already registered but not employed for any purpose. They may then be purchased 'off the shelf' for a relatively modest fee. Of course, the vast majority of these are bought for totally legitimate purposes, but the point here is that launderers may acquire a fully registered company with ease.

To all this should be added the ubiquitous trust arrangement prevalent in common law jurisdictions, which can also be potentially abused. A trust arrangement occurs where a person (a trustee) holds property as its nominal owner for the good of one or more beneficiaries. Who those beneficiaries are may not be immediately apparent.

In the United States it is generally not necessary for a non-regulated, privately held business to provide beneficial ownership information as to its equity to a governmental body. Nor does the public have access to information concerning the ownership of private firms and there is no requirement for the public filing of such information. In the United Kingdom, it is possible to hide the true beneficial ownership of companies behind nominee directors and shareholders and so the full ownership structure is sometimes opaque. This has changed somewhat with the requirement of a Persons of Significant Control (PSC) Register, which came into force in April 2016 and seeks to register all persons and corporate entities with significant control (in addition to existing registers of directors, secretaries and members, and the like). While the PSC Register now requires all registered businesses covered by the legislation to keep a register of these previously concealed controllers or beneficiaries, the fact is that for years, ownership structures were far from transparent and it still remains unclear whether all the information now on the register is correct or complete.[35] It should also be noted that the British government intends by 2021 to complement the PSC Register with a register of beneficial owners of UK property and also a register of those bidding for public procurement contracts. Furthermore, British Overseas Territories must by the end of 2020 make public beneficial ownership information of locally registered companies. Of course, it remains to be seen whether such registers will actually be successful in revealing true ownership structures in the face of concerted efforts not to divulge this kind of information.

When money is streamed through an arrangement employing any or all of these entities and when such entities are purposively bereft of detailed and transparent beneficial ownership information, the layering process can be substantially enhanced. The technical complexity of layering is what makes financial crimes so difficult for non-experts—such as juries—to grasp. Layering itself can also act as a deterrent to hard-pressed and overwhelmed anti-money laundering professionals whose job it is to try to unravel the underlying connections.

The layering process of the money launderer is not only abetted by instruments and legal entities but also by jurisdictions where regulatory and compliance requirements are often not as transparent as they perhaps should be. Certain jurisdictions, including several UK Crown Dependencies, have historically attracted business by permitting beneficial ownership records to be kept secret. While as noted there has been a move to make such information available to law enforcement agencies on request, this remains information that is generally not publicly available. There may be entirely legitimate reasons for persons wishing to keep their identities secret, but such secrecy obviously has major attractions for those who have illegal objectives.

The task of catching the launderer becomes even more difficult following the integration process. While it is often difficult to distinguish between the final layering stage and integration, the latter essentially involves delivering the funds out of the complex series of financial transactions in which they have been layered, into the legitimate economy in a fashion which makes it difficult to differentiate those originally illegal funds from the mass of legitimate funds within the mainstream legal economy. The launderer will use the funds to purchase assets such as real estate, existing legal companies or luxury goods. By so doing the launderer now also has the opportunity of increasing his wealth as the assets themselves increase in value. At the same time the launderer benefits from the employment and enjoyment of those assets.

A very good example of money-laundering integration is the London property market. It has witnessed an unprecedented boom in demand and resulting high prices but has also been the object of attention of individuals (especially those with significant political roles and connections) and entities based in several offshore jurisdictions seeking to launder money into the British economy. Property is an important integration target as it can store large amounts of capital, potentially provides significant capital gain, and adds to the good life by enhancing criminal lifestyle quality. Residential property is generally recognised to be a greater risk than commercial property when it comes to money laundering because of its turnover potential, though commercial property too is problematic as it can serve as a locus for cash-intensive businesses connected to predicate offences or money laundering itself.

In 2016, the Home Affairs Select Committee concluded that the London property market was the primary avenue for the laundering in the UK economy of £100bn of illicit money a year. It noted that poor supervision and enforcement in the London property investment market was making it a safe haven for laundering the proceeds of crime. Mr Keith Vaz, MP, chair of the Committee, claimed:

> At least a hundred billion pounds, equivalent to the GDP of Ukraine, is being laundered through the UK every year. ... London is a centre for money laundering, and its standing as a global financial centre is dependent on proactively and effectively tackling money laundering. Investment in London properties is a major route which tarnishes the image of the capital. Supervision of the property market is totally inadequate, and poor enforcement has laid out a welcome mat for launderers and organised criminals.[36]

The evidence upon which the Committee based its analysis was certainly compelling. In a submission before the Home Affairs Committee, Robert Barrington, the executive director of Transparency International UK, argued that the United Kingdom was an attractive centre for money laundering because of historical connections with the Overseas Territories and Crown Dependencies. These connections meant that it was possible to move money extremely rapidly to and from those jurisdictions not least because banks, lawyers and accountants 'will have very close connections [with those jurisdictions] and can easily set up shell companies and so on'.[37]

Sobering research by Transparency International UK clearly supports these assertions and demonstrates that London remains an extremely popular choice for money launderers who employ opaque corporate vehicles based in jurisdictions such as Panama, Jersey and the British Virgin Islands to purchase property on behalf of hidden beneficiaries. Indeed, London property is a major international target for persons laundering the proceeds of their corruption:

> Corrupt individuals may choose London property for a range of reasons, from its location as a global hub to the air of respectability it gives them. However, the property market is particularly attractive because high property prices allow corrupt individuals to launder large sums of money within a single purchase and there are loopholes in the checks undertaken on buyers' source of wealth.[38]

Transparency International's research goes on to argue that ownership information was not accessible for almost 50 per cent of overseas companies owning London land or property. Nor was any data available on the actual owners of more than half of the 44,022 land titles held by overseas companies in London. In all, 91 per cent of overseas companies that owned London property did so through offshore jurisdictions, with Panama being the most popular, followed by the British Virgin Islands. In addition, nearly 1,000 land titles in London (and no doubt potentially many more) were owned by entities that may have connections to politically exposed persons, a subject we shall shortly review. Using available data, Transparency International UK has determined that the average value of land titles held by these companies was a not insubstantial £1.9 million (far higher than the average price of London property), with the most expensive property being worth over £86 million. Unsurprisingly, the majority of the land titles owned by companies whose ownership was anonymous were in prime central London locations with 3 per cent of these being in the City of Westminster, 16 per cent in Kensington and Chelsea, and 5 per cent in Camden.[39]

The thrust of the Transparency International UK study has also been backed up by research carried out by the BBC, which uncovered that as of January 2018 there were approximately 97,000 properties in England and Wales owned by overseas firms of which 23,000 are owned by entities registered in the British Virgin Islands. In terms of ownership of these overseas properties, the British Virgin Islands is, according to this investigation, the most popular offshore jurisdiction, followed by Jersey, Guernsey and the Isle of Man with others to be found in Hong Kong, Panama and the Republic of Ireland.[40] The concern of this report was simply tax evasion—as noted above, a key purpose of money laundering—but clearly the anonymity offered by using offshore vehicles in these jurisdictions means that general money laundering by politically exposed persons and others cannot be ruled out.[41] Donald Toon, a former head of the National Crime Agency, admitted the size of the problem when he was quoted as saying that 'the London property market has been skewed by laundered money. Prices are being artificially driven up by overseas criminals who want to sequester their assets here in the UK.'[42]

Of course, in terms of integration, property may be the most visible product but it is not the only one in which money launderers integrate their finances. Luxury assets such as works of art, jewellery, cars and yachts may serve integration purposes. Organisations dealing with these are defined by the relevant UK money laundering regulations as high value dealers (HVDs).[43] In general terms, a business will be defined as an HVD where it transacts in goods and receives or makes cash payments equal to €10,000 or more in any currency whether by way of a single or set of transactions.

The key attraction of the products sold by HVDs is their portability. Property, while capable of absorbing large sums, increasing in value and providing lifestyle options for the purchaser, cannot be moved across borders. On the other hand, jewellery and yachts and private aircraft can be more or less readily transferred between jurisdictions. Her Majesty's Revenue and Customs (HMRC) divides the HVD sector into 25 subsectors of which the greatest risk areas are determined to be motor vehicles, jewellery and alcohol. While alcohol may seem a rather odd subsector to launder funds, *NRA 2017* employs a case study in which a wholesale alcohol dealer used its high value dealer status to launder funds. The alcohol dealer's bank account received approximately £3 million funnelled through various branches. Ultimately these funds were transferred to other businesses and personal bank accounts.[44]

Despite these attractions, the United Kingdom's *NRA 2017* report viewed the HVD sector as relatively low-risk because of the limited ability of the sector to absorb large sums of money. However, the role of this sector may be increasing as laundering efforts are pushed out of other areas due to increasing regulation.

The Challenge of Politically Exposed Persons

General Sani Abacha ruled Nigeria during much of the 1990s and while in power he managed to remove $3–$4 billion from the Nigerian Central Bank and launder these funds into overseas bank accounts. Court documents indicate that the money was diverted from state coffers in several ways. $750 million was paid into Nigerian bank accounts and then laundered through various European bank accounts owned by Abacha and his associates. This was done for 'state and security purposes', or so his defenders claimed. In addition, a company of which General Abacha and his associates were beneficial owners bought bills of exchange which had been guaranteed by the Nigerian Central Bank and which were later repurchased by the Nigerian Ministry of Finance on the instructions of Abacha in a deal that left the beneficiaries in profit of DM 500 million. Finally, a British company owned by Abacha's associates sold vaccines to a Nigerian family support programme—of which General Abacha's wife was the president—at an alleged profit of $80 million. The laundered money ended up in many overseas banks including within Switzerland, the United Kingdom, France, Luxembourg and Liechtenstein.[45]

The Nigerian authorities that replaced the corrupt and disgraced Abacha regime sought to recover the money and in 2001 began to lodge complaints with several European agencies, including most significantly the Federal Office of Police of Switzerland which, in turn, launched a major inquiry into dozens of Swiss banks. Ultimately some of the money was returned to Nigeria.

It is in the context of the Abacha investigation that the concept of a politically vulnerable person distinct from a criminal and terrorist (though they may be both) gained traction. Such a person was viewed as one whose political position makes him or her a high risk in respect of money laundering, corruption and bribery as well as other financial crimes. Indeed, over the years, world leaders such as Hosni Mubarak, Muammar Gaddafi and Augusto Pinochet have allegedly used their political status and connections to launder funds across the global financial system.

Concerns about the implications of corruption and bribery for the integrity and security of the global financial system and national economies was given form in the 2003 United Nations Convention against Corruption (UNCAC), which, while not using the term 'politically exposed persons' (PEPs), notes the particular vulnerability of public and foreign officials holding legislative, executive, administrative or judicial positions to bribery and embezzlement, money laundering, misappropriation, trading in influence, abuse of functions, concealment and obstruction of justice. Article 20 of UNCAC, which deals with illicit enrichment, imputes criminal behaviour to individuals whose assets cannot be explained in relation to their lawful income.[46]

Corruption may take several forms. The focus on what have become defined (according to several anti-money laundering and anti-corruption bodies) as PEPs concerns primarily so-called 'grand-corruption', which covers 'acts committed at a high level of government that distort policies or the central functioning of the state, enabling leaders to benefit at the expense of the public good'.[47] According to Transparency International's Corruption Perceptions Index 2017, which ranks countries according to their perceived risk of public sector corruption, the worst countries for corruption included (starting with the worst) Somalia, South Sudan, Syria, Afghanistan, Yemen, Sudan, Libya, North Korea, Guinea-Bissau, Equatorial Guinea, Iraq, Turkmenistan, Angola, Eritrea, Chad and Tajikistan.[48]

One form of corruption inextricably linked to PEPs is bribery. Bribery can be defined as the offer and acceptance of an inducement for an action which is illegal or unethical. Inducements include gifts, fees and other advantages. It enriches persons at the expense of the institutions where they are employed, and distorts markets and the allocation of resources. The international financial, defence and construction industries are sectors known for bribery vulnerabilities in respect of PEPs, especially when those activities take place in jurisdictions with generally high levels of corruption.

While such corruption and bribery activities may take place in overseas jurisdictions, the foreign investor will not be immune because his own representatives may be involved in the so-called supply side of corruption and bribery in terms of actually paying bribes or being involved in other aspects of corruption for purposes of facilitating deals. Moreover, foreign corrupt officials, including PEPs, may attempt to launder their ill-gotten gains out of their states and back into the legitimate economy based in any number of international financial centres.

There is no agreed definition of what constitutes a PEP but the general consensus is that PEPs may abuse their power for personal gain or for the benefit of family members and associates, conceal funds secured by corruption or bribery (including by way of laundering the funds), and gain access to and control of legal entities for similar purposes. PEPs are often viewed as including:

- heads of state, government and ministers;
- senior judicial officials whose decisions are not subject to appeal;
- senior officers in the armed forces;
- members of royal families with governing responsibilities;
- senior executives in economically or politically important state-owned enterprises;
- senior officials of major political parties;
- close family members and associates of the above.[49] ⟵

Whether such a characterisation is fair is debatable. Why such persons should attract increased scrutiny by financial and other institutions and effectively bear an assumption of guilt is not an argument we will address here. What we can note is that when dealing with these persons, public authorities and private institutions are required to presume potential malfeasance and undertake a higher level of due diligence of the type we shall review in Chapter 5. That having been said, concern about PEPs was not something that overly troubled financial institutions in the past.

In a 2011 study by the United Kingdom Financial Services Authority (FSA) it was shown that numbers of banks were hesitant in turning away or withdrawing from very profitable business relationships even when there was PEP involvement. Around a third of banks, including private banking arms of some major banks, were ready to accept high levels of money-laundering risks as a part of such business. Furthermore, over half of banks failed to properly implement the required enhanced due diligence in respect of high-risk PEP situations. The FSA found that a third of banks even failed to set in place measures aimed at identifying customers as PEPs. Three-quarters of banks failed to take adequate measures to establish the source of wealth and funds used in the business relationship. In a third of banks the management of customer due diligence records was inadequate and banks were unable to provide an adequate overview of their PEP relationships.[50]

Since 2011, relevant due diligence has improved not least because of the growing awareness of the problem by banks, partly driven by the threat of fines and fear of reputational damage and also due to improvements in technology helping financial institutions identify PEPs. For example, LexisNexis has a global PEP list, while Thomson Reuters, Dow and others also provide searchable global PEP databases that can support due diligence.

Nevertheless, despite all these due diligence improvements, the issue of PEPs still continues to plague the United Kingdom's banking system. This was revealed after the poisoning of the former security agent Sergei Skripal and his daughter Yulia in Salisbury in 2018, allegedly by Russian government agents. The fallout from that affair turned attention to ways of punishing President Putin by exerting pressure on Russian PEPs who had invested significant sums of money in the United Kingdom, including, according to some, integrating laundered money into the London property market. A House of Commons Foreign Affairs Committee investigation into Russian PEPs active in the UK received evidence that 'London [remains] a "top destination" for Russian oligarchs with links to the Kremlin to launder proceeds of corruption and to hide their assets.'[51] Information presented to the committee suggested that over 20 per cent of the 176 properties worth £4.4 billion that had been purchased with what is termed 'suspicious wealth' involved Russian individuals.[52]

A reading of the House of Commons report indicates that part of the issue is not simply a matter of defective due diligence on the part of the private sector (though the view is that some active in this sector may sometimes be willing to turn a blind eye to the origin of funds) but policy decisions of the state which amount to toleration of such PEP activity. Whatever the case, the implications of permitting PEP money-laundering activity for national security in general and the financial sector in particular are clear. The report is emphatic:

> The use of London as a base for the corrupt assets of Kremlin-connected individuals is now clearly linked to a wider Russian strategy and has implications for our national security. Combating it should be a major UK foreign policy

priority. The assets stored and laundered in London both directly and indirectly support President Putin's campaign to subvert the international rules-based system, undermine our allies, and erode the mutually-reinforcing international networks that support UK foreign policy. The size of London's financial markets and their importance to Russian investors gives the UK considerable leverage over the Kremlin. But turning a blind eye to London's role in hiding the proceeds of Kremlin-connected corruption risks signalling that the UK is not serious about confronting the full spectrum of President Putin's offensive measures.[53]

Key Sectors Vulnerable to Money Laundering

No matter the jurisdiction in question, there are several sectors within the modern global economy that present particular vulnerabilities to the criminal money-laundering methodologies that we have reviewed. In addition to the HVDs, these include:

- the banking sector;
- the professional services sectors such as legal services, accountancy services, and property and estate agency services;
- the gambling sector;
- money service businesses;
- non-profit organisations; and
- the financial technology sector.

While obviously this is not a conclusive list, and the relative importance of these sectors may vary depending upon jurisdiction, those identified here because of their high cash flow, their gateway role or their means of transferring funds are all, in the context of placement, layering and integration, certainly potentially vulnerable to exploitation by money launderers. Here we shall look in greater detail at the banking sector and the professional services sector.

Banking and Financial Services

At the core of any modern financial system is the banking sector. Banks provide a wide range of financial services such as accepting deposits, making loans, exchanging currency and managing wealth. One set of banks—commercial or retail banks—provide day-to-day services to individuals and businesses such as making loans, providing payment services and accepting deposits. Investment banks—sometimes described as those working in the wholesale sector—deal primarily with corporate clients and are involved in providing sophisticated commercial products and services. These may include support in terms of underwriting, merger and acquisition activity, and hedging services. Given its systemic centrality (some consider it 'too big to fail'), the banking sector is one of the most regulated sectors in any economy.

Some of the key vulnerabilities of the banking system to money laundering can be located in the following areas:

- structuring of deposits in the placement phase of money laundering so as to avoid reporting and recording requirements;
- misuse of legal entities such as shell or shelf companies and the disguising of ownership structures by, for example, using accounts in someone else's name or the name of a business;
- schemes such as trade-based money laundering and complicit brokers and other insiders;
- exploitation of various payment technologies; and
- third-party payment processors (for example such as PayPal) which facilitate retail transactions but could be targeted by criminals to launder funds.

Moreover, there are vulnerabilities in respect of correspondent banking relationships, which we shall deal with in the context of fund movement, and of course general compliance deficiencies, which we shall address in greater detail in Chapter 4.

In the United Kingdom, for example, the general banking sector has been identified as being at high risk to money laundering, with retail banks exposed to the greatest volume of criminal activity of any financial sector. As *NRA 2017* put it: 'While controls are more developed in retail banking than other areas, the widespread criminal intent to exploit retail banking products and the increasing speed and volume of transactions means that the sector remains at high risk of money laundering.'[54]

The fact that funds can be withdrawn in cash and readily transferred overseas makes retail banking highly attractive to launderers both at the high and low ends. The speed and volume of transactions, especially those in electronic form, are key to understanding the retail banking industry's vulnerability because these factors make oversight extremely challenging.[55] High-end money laundering of the type identified above often takes place outside the retail banking sector and within the wholesale banking space where larger quantities of funds can move and more readily be placed, layered and integrated. As noted, services provided by international investment banks include high-volume debt and equity services and currency transactions. These transactions become attractive to money launderers especially when executed across jurisdictions by way of branches, subsidiaries or correspondent banking relationships.

A good example of relevant vulnerabilities to money-laundering schemes within the banking sector is the problem Deutsche Bank encountered in Moscow in 2015 when it found itself engaged in 'mirror trades' and ended up paying nearly $700 million in penalties to the US and British authorities. A mirror trade is characterised by two stock transactions in separate locations in which money is moved out of one of the locations and into the other. This of course is not in itself inherently illegal but it is a trade open to corruption. Between 2011 and 2015 Deutsche Bank, acting through its Moscow subsidiary, supported a Russian company using roubles to purchase blue-chip Russian shares such as Lukoil. At about the same time Deutsche Bank, acting in London on behalf of an offshore company, sold the same shares for dollars, euros or pounds and, by doing so, effectively converted roubles in Russia into foreign currency outside that country with part of the proceeds ending up in accounts in Cyprus and the Baltic states. Approximately $10 billion was moved out of Russia in this manner.[56] As explained by investigative journalist Ed Caesar writing in the *New Yorker*: 'Both the Russian company and the offshore company had the same owner. Deutsche Bank was helping the client to buy and sell to himself.'[57]

The Deutsche Bank mirror trade case is just a small example of the amount of high-end laundered money passing through the wholesale banking sector. In 2017, the *Guardian* revealed what it called the 'Global Laundromat' in which between 2010 and 2014 at least $20 billion was laundered by various methods out of Russia—the actual figure was potentially as high as $89 billion. According to the report, British-based banks processed nearly $740 million of these laundered funds while British registered companies also played a significant role as did firms based in offshore jurisdictions.[58] The documents refer to about 70,000 banking transactions, including 1,920 that passed through UK banks and 373 via US banks. According to the *Guardian*: 'HSBC, the Royal Bank of Scotland, Lloyds, Barclays and Coutts are among 17 banks based in the UK, or with branches here, that are facing questions over what they knew about the international scheme and why they did not turn away suspicious money transfers.'[59]

Professional Gateway Services

Another key sector that remains vulnerable to money laundering is professionals working in the financial arena such as lawyers, accountants and estate agents. Part of the vulnerability here stems from the fact that these professionals act as gatekeepers to the financial system, enabling access to accounts, products and assets while also, in the case of lawyers and accountants, by their participation, conferring legitimacy, intentionally or inadvertently, on the laundering process.

Specific vulnerabilities in the case of lawyers and accountants include coerced professionals who are targeted by criminal elements, colluding professionals who knowingly work with criminals to place, layer or integrate illicit funds, and negligent professionals who fail to undertake proper due diligence of their clients and their plans and do not make the proper reporting to the relevant authorities.[60] It is the negligent type (or those that prefer to be wilfully blind) who form the majority of vulnerable professionals.

In respect of accountants and lawyers, among their greatest attractions to money launderers is the appearance of respectability that their involvement gives to the money-laundering transaction. Hiding behind accountants and lawyers is a strategy that criminal money launderers have long adopted. In terms of actual services, company formation and the structuring of accounts are what the launderers are often seeking. An *NRA 2017* case study uses the example of an accountant who together with a Russian national acted as joint director of a UK-registered company. The accountant established structures that transferred more than $60 million through Russia, Cyprus, the Czech Republic, Latvia and the British Virgin Islands. Ostensibly, the company offered high-end leisure services, but actually it served to launder funds owned by sanctioned individuals out of Russia.[61]

Accountants and lawyers provide not only trust and company formation facilities but also insolvency services, which similarly can act to break linkages between the origin of funds and their present ownership and use. False accounting in both the high end and cash-based end of the market similarly acts to disguise the origins and purpose of funds. The use of accountants' client accounts to launder money is also problematic. Launderers with no evident business rationale can place their illegal funds into their accountant's client account or accounts and then retrieve it at a later date or move it on, thereby distancing themselves from the actual origin of the money. And there is always the possibility of accountancy services being used for tax evasion and other tax frauds.

Lawyers' roles in conveyancing represent another primary risk area in this sector, especially when linked to trust formation aimed at disguising beneficial ownership and complicating the due diligence process. The mismanagement of client accounts, again in respect of legal property services and the transfer of funds to third parties, presents another problem. The UK National Crime Agency reports that criminal launderers may deliberately compartmentalise legal work between or within firms to add to layering and to complicate due diligence. The Solicitors Regulation Authority (SRA) in its 2017/18 Risk Outlook has noted that the number of money laundering-related matters reported to it increased by nearly 20 per cent in 2015/16, and has remained static at around 175 per year ever since. One-third of all anti-money laundering issues reported to the SRA over a three-year period related to residential conveyancing, three times more than any other field of law.[62]

Estate agents can be involved in the assistance they provide criminal money launderers in purchasing and selling property, and property as we have continually noted throughout this chapter represents both a key money-laundering medium and target. Surprisingly, however, the *NRA 2017* report does not place too much emphasis on the significance of estate agents' involvement in the laundering process as 'much of the risk lies with those closer to the client and their funds, such as legal professionals'.[63]

Within the literature there has emerged a new concept of professional money laundering, which may involve any of the abovementioned professionals but which is more particularly defined as consisting of individuals, organisations or networks who for a fee or a commission will launder the proceeds of crime.[64] Such persons or organisations are usually not involved in the original predicate offence itself but work to provide expertise so as to disguise the source of the funds, the controlling interests and the destination. An example of this business model is that of the now defunct Khanani money-laundering organisation. This enterprise was supervised by a Pakistani national, Altaf Khanani, who acted for a range of Colombian, Mexican and Chinese criminal groups and laundered billions of dollars between the United Arab Emirates, Pakistan, the United States, the United Kingdom, Canada, Australia and several other jurisdictions. The laundering of funds was also undertaken for several terrorist groups including the Taliban, Lashkar-e-Taiba, Dawood Ibrahim and al-Qaeda. Drug proceeds were deposited via multiple wire transfers from several general trading companies in exchange for a three per cent commission of funds laundered. In 2016 Khanani pleaded guilty to US federal money-laundering charges and in 2017 was sentenced to 68 months' imprisonment for conspiracy to commit money laundering.[65]

We began with a list of famous movies and TV series depicting drug and other illicit activities and the theme of money laundering that permeates many of them. Perhaps the greatest movie series depicting organised crime is *The Godfather*, directed by Francis Ford Coppola and based on the book by Mario Puzo which tells the story of the Corleone family's rise to prominence, first in rural Italy, then in the United States, and finally in Italy again. Interestingly, this story too has a permeating money-laundering theme that reflects the effort (ultimately one that at least partly fails) by organised crime to break the link to its original sins and enter the legitimate business world.

So in *Godfather I* we witness the rise of the Corleone family mainly in New York. Initially involved in small-time crime such as robbery and protection rackets, they relentlessly expand into more weighty and lucrative businesses, an expansion that is inevitably accompanied by the violent neutralisation of competing criminal organisations. When Vito Corleone (Marlon Brando) refuses to enter the drug trade, we are not sure why, but his disappointment when his son Michael (Al Pacino) is unable to distance himself from his criminal roots is visible.

In *Godfather II*, the family moves out of its New York confines to Las Vegas. The gambling industry holds several attractions for the Corleone family. First, it is a legitimate business unlike their New York-based activities. Second, it is itself a highly lucrative business with the potential of significant profits. Finally, it is a cash-intensive business ideal for the integration and layering stages of the laundering process. A secondary storyline is Michael Corleone's investigation of options for investment in pre-revolutionary Cuba. The lack of regulation and oversight in that chaotic and corrupt jurisdiction also held promise for laundering funds originally generated in the United States, for also realising profits within Cuba, and then for repatriating such funds—now seemingly legal—back into the United States.

In the final Godfather movie, *Godfather III*, the objective of transforming the Corleone business into a legitimate enterprise is explicit. The movie in fact starts with Michael Corleone establishing a charity (no doubt seeded with essentially criminal funds), something which in itself bestows legitimacy, while at the same time potentially serving as a vehicle for laundering funds. However, the main plot line of the film revolves around an attempt to purchase a company called Internazionale Immobiliare—a real estate holding company owned by the Vatican. Real estate is an obvious major placement vehicle for any money-laundering plan and would have served the Corleones well in breaking the link with their funds' origins. But perhaps most important would be the fact that the business was previously owned and connected with the Vatican, a source of both temporal and spiritual legitimacy. The Corleones' transformation into the realm of legitimacy would have been significant, to say the least. Their money would have started in small-time criminality within Italy but would ultimately have found a resting place within the Holy See.

Concern about criminal money laundering has existed for decades and, as we shall note in later chapters, has given rise to a sophisticated and complex architecture to combat its prevalence. Before we can begin to assess those structures, we need to analyse another core threat to the financial system: its exploitation by terrorists. As we shall soon discover, the terrorist financial threat in many respects represents a very different type of challenge from criminal activity to the integrity of the financial system.

2

THE FINANCING OF TERROR

[Al-Qaeda's supporters] are aware of the cracks in the Western financial system as they are aware of the lines in their own hands.

Osama bin Laden

Terrorists and Finance

Terrorist finance can be defined as the activities of terrorists aimed at generating and transferring funds and using them for purposes of organisational maintenance and growth and, ultimately and most importantly, for the support of terrorist operations. In these activities there are continuities with the methods of criminals in terms of fund generation and fund movement. Moreover, in view of the continuities and similarities in money-laundering techniques and strategies, criminal money laundering and terrorist financing are often referred to as two sides of the same coin, with structures and processes used for identifying and countering one type of financing often employed for identifying and countering the other. However, there are also characteristics in the generation and transport of terrorist funds and in the ends to which terrorists seek to put their monies that differ markedly from the criminal sector. The example of al-Qaeda helps demonstrate both these common and unique characteristics of the terrorist financing model as well as the significance of finance in underpinning terrorist operations.

Finance, al-Qaeda and the 9/11 attacks

No terrorist operation has come close to the direct and indirect impact of the 9/11 attacks by al-Qaeda on the United States. Nearly twenty years after the airborne assaults on New York and Washington, we live amidst the consequences of that act of terrorism in terms of domestic human rights, intercommunal relations, attitudes and tolerances, foreign policies and overseas wars.

The interrogation of Khalid Sheikh Mohammed, one of the key organisers of the 9/11 attacks, revealed that the total cost of the operations against the World Trade Center and the Pentagon on 9/11 was a paltry $400,000 to $500,000. The financial consequences of these strikes, according to one estimate, has been over $3 trillion, or about $7 million for every dollar al-Qaeda spent on first planning and then executing the attacks. Specifically this estimate of economic consequences—obviously a very general estimate with enormous caveats—consists of the toll of physical damage ($55 billion); the economic impact ($123 billion); homeland security and related costs ($589 billion); war funding and related costs ($1,649 billion); and future war and future veterans' care costs (($867 billion).[1]

While at the public level we only have a generalised picture of al-Qaeda's financial history, it seems evident that in setting up his global organisation Osama bin Laden brought with him an understanding of financial networks, fund generation and fund disbursement. Indeed, when the US Central Intelligence Agency (CIA) first began investigating him he was viewed primarily as an international terrorist financier.

Much of this financial knowledge must have stemmed from his involvement in or at least proximity to his family's construction business in Saudi Arabia—a firm that still exists and functions. The development of the Bin Laden organisation began in the 1930s during the Saudi Kingdom's early years, under the reign of King Abdul Aziz Al Saud. Mohammed bin Laden, father of Osama and also of Osama's half-brother, Bakr bin Mohammed bin Laden, the later chairman of the Saudi Binladin Group,[2] in 1931 founded the Mohammed bin Laden business as a general contracting company. From then on, the group's history and growth became very much entangled with the history and politics of Saudi Arabia itself. In 1950, when King Abdul Aziz Al Saud, who had founded the Saudi Arabian Kingdom in 1932, was planning the first Saudi extension to the Prophet's Mosque in Medina, he asked Mohammed bin Laden to undertake this sensitive and important work on behalf of the Kingdom. Construction of the extension was ultimately completed during the reign of King Abdul Aziz's son, King Saud bin Abdulaziz.

As a result of the success of this project, in the mid-1950s Mohammed bin Laden was requested to extend the Holy Mosque in Makkah, another sensitive piece of work as it was the first such extensive project to be initiated for hundreds of years. Construction continued through the reign of Saud's successor King Faisal, and was ultimately finished two decades later during the reign of King Khalid. In 1964 Mohammed bin Laden was employed to undertake construction at the Dome of the Rock in Jerusalem, where his firm reclad the mosque's dome, a prominent fixture on the Jerusalem skyline. According to the writer and Middle East analyst John Cooley, Mohammed bin Laden 'once had the unique experience of saying, in one day, morning prayers in East Jerusalem (before Israel's 1967 conquest of the city), noon prayers in Medina and evening prayers in Mecca'.[3]

When Saudi Arabia moved on to upgrade its rudimentary infrastructure, Mohammed bin Laden's business expanded. The company became involved in several important national projects including the construction of a network of highways which traversed the breadth of the country, enhancing its trade and communications potential. Furthermore, as Saudi Arabia developed and extended its regional and global reach, the Bin Laden business, now under the leadership of Salem M. bin Laden, Osama's late half-brother, expanded together with it. The company's activities diversified from roadworks and construction to include airports in Fujairah, Damascus, Cairo and Aden and hotels in Jordan and Kuala Lumpur. Organisationally, so as to better adapt and function within a global context, the Saudi Binladin Group was set up to incorporate various Bin Laden companies under a single parent.[4]

Osama grew up within a sophisticated family business environment and gained some kind of experience of and insight into international financial matters including apparently the raising of finance and the transfer of funds. He also, alongside his other brothers, became a shareholder in the Binladin organisation. It was originally estimated that Osama inherited approximately $300 million from his father, but even before 9/11 the US authorities had become convinced that this was not the case and that rather, between 1970 and 1974, he received a stipend of approximately $1 million a year. In the 1990s the Saudi government, which had then fallen out with Osama, forced the Binladin Group to find a purchaser for Osama's share and buy him out of the business. He was thus cut out from the family organisation and distanced from it—a fact that has enabled the family business to continue functioning within Saudi Arabia and the Gulf to this day.

By the 1990s, Osama bin Laden was already firmly under the scrutiny of the US intelligence organisations. After the 1998 al-Qaeda attacks in Nairobi and

Dar es Salaam, National Security Council officials visited Saudi Arabia (in 1999 and 2000) to discuss, among other things, terrorist financing issues. During their various visits to Saudi Arabia, they met with members of the Binladin Group and began to gain a fuller understanding of Osama's financial background.[5] Already in this period US intelligence agencies' interest in Osama was significantly financial. For example, classified studies referred to in the 9/11 Commission Report prepared by the National Commission on Terrorist Attacks upon the United States include 'Usama bin Laden Finances: Some Estimates of Wealth, Income and Expenditures' (November 1998); and 'Pursuing the bin Laden Financial Target' (April 2001).[6] Certainly, the US intelligence services prior to 9/11 were interested in Osama as a financier, and in the words of Peter Bergen '"a Gucci terrorist"—one who financed some acts of terrorism but had no larger operational role'.[7]

Almost two decades previously, in 1980, Osama had arrived in Afghanistan and soon become central in helping fund the anti-Soviet jihad. He appears to have recognised that not only did the campaign need to develop global roots for its recruitment and political support but also critically for its financial backing. Osama played a pivotal role in transforming the war against the Soviet Union in Afghanistan into a global jihad backed by global financial resources. The US 9/11 Commission Report estimates that it cost al-Qaeda approximately $30 million a year to sustain itself in terms of maintenance and operations. This money came primarily from Gulf-based donations and charities.

Indeed, to financially underwrite the Afghanistan military campaign Osama helped develop what became known as the 'Golden Chain'—a support network backed by very wealthy and ideologically committed individual patrons in the Gulf States and Saudi Arabia who were willing to contribute significant private donations to al-Qaeda.[8] Some of these individuals knew where the money was going but possibly some did not. Some no doubt wilfully turned a blind eye. This financial support network was broader than simply a group of high net-worth individuals. According to the 9/11 Commission Report, the support base extended to religious leaders in various mosques and financial facilitators and relied quite heavily on certain imams who were willing to divert *zakat*—charitable donations made by congregants as part of their religious duties—to al-Qaeda.

Charities became key to Osama's global financial network. Al-Qaeda adopted two approaches to charities. First, the organisation relied on sympathisers within large international charities with various foreign branches and lax oversight mechanisms to syphon money to al-Qaeda accounts. One example of this was the Saudi Arabian-based al-Haramain Islamic Foundation. The large charities also served to support smaller charities in different parts of the world and these too proved sympathetic to those prepared to divert funds in favour of al-Qaeda causes. A second approach was to control the charity itself—the al-Wafa organisation being an example. Here the charity provided both cover for al-Qaeda agents and a front company for raising and distributing al-Qaeda funds.

During the jihad in Afghanistan, the United States and Saudi Arabia actively provided finance to anti-Soviet fighters through the intermediary services of the Pakistani Inter-Services Intelligence Directorate. The concern was that at least some of this money was being diverted to Osama's fighters— thus giving credence to the argument that during the 1980s the United States financially supported Osama bin Laden. However, this analysis is directly disputed by the 9/11 Commission Report which states that this money did not go to Osama bin Laden's group, whose fighters relied throughout on their own sources of funding.[9]

In 1988 the Soviet Union started to withdraw its forces from Afghanistan but Osama retained in place a recruitment, organisational and financial network the core of which was based in the Arabian Gulf. He had helped create it and he certainly did not want to allow it to unravel as it could be used at a later date for other operations outside Afghanistan. During that same year he formally established the al-Qaeda organisation.

With the war in Afghanistan against the Soviet Union over, Osama could focus his radicalism on other pressing political issues. He tried to encourage the Saudi Arabian government to rid itself of American forces stationed in the country; consequently relations between him and the government in Riyadh deteriorated. As a result his position in the Binladin Group and his access to funds and the company's financial assets also began to weaken. The Saudi Arabian authorities exerted great pressure on the Binladin Group to distance itself from Osama and in 1989 Osama's fighters moved to Sudan, where Hassan al-Turabi, the Sudanese leader, provided protection for the organisation. But the benefit of its presence worked in both ways as al-Qaeda came to provide critical financial support to the hard-pressed Khartoum government. Sudan also served as a hub for Osama's global financial network.

While in Sudan, Osama, who had himself moved to Khartoum in 1991, began financially and otherwise to support the Sudanese government in its war with the Christian and animist south. At the same time, he also engaged in construction projects, with which he had experience from his family business. As the 9/11 Commission Report notes, while in Sudan Osama 'set up a large and complex set of intertwined business and terrorist enterprises. In time, the former would encompass numerous companies and a global network of bank accounts and non-governmental institutions.'[10] It should also be recognised that elsewhere the official US analysis somewhat contradicts itself when it emphasises the point that Osama remained dependent on external donations and that most of the Sudanese businesses which he operated between 1991 and 1996 'could not have provided significant income, as most were small or not economically viable'.[11]

Al-Qaeda soon began helping the Sudanese government build a new highway that stretched from Khartoum to Port Sudan on the Red Sea as well as being involved in other construction projects. Osama also established the Wadi al-Aqiq company, which served as the parent company for several local businesses.[12] Other enterprises included an equipment leasing business, a tanning factory, a bakery, an investment firm called Taba, a cattle-breeding company, and a fruit and vegetable import and export firm.[13] Osama also involved himself in banking activities, according to some reports, by helping capitalise the Shamal Islamic Bank by as much as $50 million, though this amount may be somewhat overstated. The bank was able to provide al-Qaeda with a set of global correspondent banking relations that gave the group an international financial reach. In addition, he also became active in the informal banking sector through establishing money transfer systems that worked out of Sudan.

According to the 9/11 Commission Report, Osama used his base in Khartoum to create a global business empire in support of terrorist operations. He set up a major business enterprise in Cyprus, a branch in Zagreb, a charitable foundation in Sarajevo which supported Bosnian Muslims fighting in Serbia and Croatia, and an NGO in Baku, Azerbaijan, that was used by Egyptian Islamic jihadists as a means of transferring monies to Muslim rebels in Chechnya. He also created a charitable organisation, the Third World Relief Agency, which was based in Vienna but had branches in Zagreb and Budapest. Later another NGO was set up in Nairobi as a cover for operatives working in East Africa and it helped support the 1998 attack in Nairobi and Dar es Salaam, operations that may have amounted to $10,000, a not insignificant sum given the challenges of financial transfer. As Lawrence Wright has noted, in an impoverished country such as Sudan, Osama's finances amounted to a second economy.[14] Financial capability more than fighting capability made al-Qaeda attractive to the Sudanese regime.

Financial resources also underpinned his building of alliances throughout not only the Middle East and Central Asia but also Asia, where he provided support for the Abu Sayyaf Brigade in the Philippines and Jamal Islamiyah in Indonesia. In the United States, a non-profit organisation called al-Khifa was set up in several cities including Pittsburgh, Atlanta, Boston and others. These branches served to raise funds for Osama's projects. Al-Qaeda operatives also appear to have set up procurement networks in Western Europe and Asia.[15] It was ultimately this global financial network that gave weight to his plans for a global terrorist network.

During this period Osama's relations with the Saudi government, already under enormous strain because of his opposition to the 1991 war with Iraq and the resulting presence of American troops in Saudi Arabia, finally ruptured. In 1994 the Saudi Arabian government took away his passport and froze all his assets and required the Binladin organisation to find a buyer for Osama's share of the company. The proceeds of this sale were initially frozen and Osama never appears to have gained access to them.[16] These developments certainly support the assertion that ultimately his personal inheritance and funds played a marginal role in al-Qaeda financing and that the organisation was supported primarily by donations.

For al-Qaeda, things were soon to get far worse. By 1996 Osama lost his base in Sudan, which came under growing US, Saudi Arabian and international pressure to deport him from the country. At the same time Osama appears to have been encountering growing financial difficulties[17]—both because of problems within the Sudanese economy and his businesses and because the Saudi Arabians leant on the Sudanese. As a result, the Sudanese government ultimately expropriated his businesses and put them up for sale. Osama and his group were once again forced to flee.

Osama bin Laden then returned to Afghanistan. The 9/11 Commission Report argues that during this second sojourn in Afghanistan he received no (or very little) support from either his personal fortune located in Saudi Arabia (from which he was cut off due to measures taken by the government in Riyadh) or his business interests in Sudan (which had essentially been confiscated). Rather it was the Golden Chain charitable network together with other charities, such as al-Wafa, which diverted funds to him or which were actual fronts for al-Qaeda, that financially supported the group.

During this second sojourn in Afghanistan, Osama, in the same way in which he had supported the Turabi government, used his financial largesse to support and ingratiate himself with the Taliban regime. The 9/11 Commission Report states that 'Bin Laden eventually enjoyed a strong financial position in Afghanistan thanks to Saudi Arabian and other financiers associated with the

Golden Chain'.[18] He was certainly involved in supplying finance to the Taliban, some of which derived from the Golden Chain charitable fundraising network in the Gulf. While in Afghanistan there is no evidence of al-Qaeda relying in any significant way upon the drugs trade as this source of revenue fell under Taliban and not al-Qaeda control. Nor, as some later reports tended to suggest, did al-Qaeda become involved in the West African blood diamonds network as a source of funds.

As to fund movement, while in Afghanistan Osama relied heavily on alternative remittance systems to secure the funds from relevant donors and charities. The financial transfer system of Hawala in Pakistan, Dubai and elsewhere in the Middle East was employed to move funds to points where they could be accessed by al-Qaeda. In any event, there was no formal banking system in Afghanistan, and after the 1998 attacks in Nairobi and Dar es Salaam Osama was being intensively monitored by US intelligence agencies and other authorities. The physical movement of cash also appears to have played a part: al-Qaeda employed the Afghan state-owned Ariana Airlines—'al-Qaeda's direct lifeline to the outside world'[19]—to courier cash into the country.

Consequently on the eve of the 9/11 attacks, according to some estimates, al-Qaeda had an annual budget of approximately $30 million, most of which was spent on maintaining the organisation including expenses for salaries, training, vehicles, arms, recruitment, propaganda and selected start-up money for various jihadist groups within Afghanistan while little was actually spent on direct terrorist operations. $10–$20 million per annum was paid by al-Qaeda to the Taliban so as to ensure its safe haven. Al-Qaeda's funds enabled the establishment of large training camps within Afghanistan; it is estimated that 10,000 to 20,000 fighters passed through these camps and benefited from al-Qaeda financing. Costs were not insignificant and the 9/11 Commission Report states: 'Before 9/11, al-Qaeda spent funds as quickly as it received them.'[20]

After 9/11 there was significant speculation—no doubt at least in part generated by Osama's background and the fact that the majority of the hijackers were Saudi Arabian—that al-Qaeda had received state support for its activities especially from Saudi Arabia. The 9/11 Commission Report was explicit in stating that aside from the Taliban no government directly supported al-Qaeda. There was no specific indication that the Saudi Arabian government (or Saudi Arabian government or royal family individuals, such as Princess Haifa al–Faisal, who at one point fell under suspicion) provided any financial backing, though the report does say that some charities with heavy Saudi Arabian government involvement could have supported the organisation. During his interrogation—which included waterboarding—Khalid Sheikh Mohammed claimed that 85 to 95 per cent of the cost of the 9/11 attacks came from Osama's personal funds, though elsewhere he told his interrogators that when Osama arrived back in Afghanistan from Sudan he had no funds at all.[21] Why he should have held the view that the origins of the funds were personal is far from clear. One explanation could be that he was concealing their source (probably donations to al-Qaeda). Another explanation—possibly a stronger one—is that he was not aware of all the actual sources of al-Qaeda funding, which would suggest a significant degree of compartmentalisation of financial information within the al-Qaeda organisation.

In addition, the official investigation into the 9/11 attacks found no evidence that advance knowledge of the attacks had allowed al-Qaeda to benefit in any way by speculating on the stock exchange as all pre-9/11 unusual trading activity was proved to have innocuous explanations.[22] So, for example, while the volume of put options (which pay off when the relevant stock price drops) surged in respect of the parent companies of United Airlines on 6 September 2001 and in American Airlines shortly thereafter, these were found to be connected to the purchase of such puts by a US-based institutional investor as part of an investment strategy and the circulation of a US-based options trade newsletter on 10 September that recommended such purchases. Both the Federal Bureau of Investigation and the Securities and Exchange Commission invested enormous efforts investigating this topic, including securing the co-operation of various foreign governments, but no linkage to al-Qaeda was proven.[23]

The 9/11 Commission Report admits that 'To date, the U.S. government has not been able to determine the origin of the money used for the 9/11 attacks'.[24] What we do know is that the monies were provided by Osama and sent to Khalid Sheik Mohammed, who then provided funds for the various hijackers to get to the United States, to train for the operation (including flight school) and live there until the attacks. Although al-Qaeda experienced funding shortfalls as part of its fundraising cycle (most money it received came during Ramadan), the report says there is no evidence that any terrorist operations were interrupted as a result of financial difficulties.

Certain of the important operatives such as Khalid Sheikh Mohammed received large amounts of cash from al-Qaeda, but most of the money that passed from al-Qaeda to the hijackers involved more modest amounts. Prior to the attack, one group of hijackers residing in Hamburg—the so-called

Hamburg cell—received no funds from al-Qaeda but funded themselves until after 1999 when they directly joined the conspiracy. One of the hijackers, Marwan al-Shehi, was supported by a salary he earned in the armed forces of the United Arab Emirates, which sponsored his studies in Germany. The members of this cell may have each received $5,000 to return to Germany following their selection for the operation in Afghanistan. Additional funds were given to them in order to travel to the United States. Other plotters may have received up to $10,000 each.

The global banking system's anti-money laundering and counter-terrorist financing mechanisms put few real impediments in the path of the terrorists. As operational planning got under way, the nineteen hijackers were funded through wire transfers or cash from Khalid Sheikh Mohammed, which they physically carried into the United States or which they deposited into foreign accounts and then drew upon when they were in the country. There does not appear to have been any use of alternative remittance systems, and where the banking system was used, it was the formal system itself. In the United States, the Hamburg operatives paid for their flight training with monies that were wired to them from Dubai by Khalid Sheikh Mohammed's nephew. Between June and September 2000, a total of $114,500 was wired to two of the hijackers in a number of transactions varying from $5,000 to $7,000. As noted in the 9/11 Commission Report, these transfers 'relied on the unremarkable nature of [the] transactions, which were essentially invisible amid the billions of dollars flowing daily across the globe' and, furthermore, the sender 'was not required to provide identification in sending this money, and the aliases he used were not questioned'.[25] While attempting to enter the United States and while there preparing for the attacks, the hijackers spent in excess of $270,000.[26] Expenses included travel to obtain passports and visas, travel to the US, the needs of facilitators and those chosen to be hijackers who did not ultimately participate in the attacks. The 9/11 Commission Report admits that 'For many of these expenses, we only have fragmentary evidence and/or unconfirmed detainee reports, and can make only a rough estimate of costs'.[27]

In the end, as noted, the total costs of the terrorist operation on 9/11 were estimated to be between $400,000 and $500,000. This figure was given to American interrogators by Khalid Sheikh Mohammed, who used the lower figure but did not include the $50,000 expenses attributable to the so-called twentieth hijacker, Zacarias Moussaoui, who did not participate in the operation. No terrorist attack before or since has cost nearly as much, but then again not many terrorist organisations could have matched the complexity of that operation, including the elements of fund generation and transfer.

In 2016 the US government released some 28 previously classified pages from its 9/11 investigation that some hoped would shed light on any possible Saudi Arabian funding connection to the 9/11 attacks. There it was noted that 'some of the September 11 hijackers, while in the United States, apparently had contacts with individuals who may be connected with the Saudi Arabian government'.[28] The pages also state: 'In their testimony, neither the CIA nor FBI witnesses were able to identify definitively the extent of Saudi Arabian support for terrorist activity globally or within the United States and the extent to which such support, if it exists, is knowing or inadvertent in nature.'[29] For those who believed that Saudi Arabia was involved in supporting the attacks, there was some information in the 28 pages that confirmed the assertion,[30] but for others the evidence was not conclusive and this release of new data never led to any theory of Saudi Arabian official support for 9/11 gaining much traction.

Following 9/11 and the destruction of the al-Qaeda central command based in Afghanistan, financing, like the organisation itself, atomised. Regional groups and individuals could no longer expect central funding support and were forced to rely primarily on locally generated sources of funds.

While the costs of the 9/11 attacks were quantitatively very different from those of any other terrorist action, this brief analysis of al-Qaeda finances around the 9/11 attacks reveals certain characteristics that manifest themselves within the general area of terrorist financing and are not unique to al-Qaeda. Indeed, the al-Qaeda example clearly reflects key elements of the terrorist financing typology. These include the financial priorities of terrorist groups and the challenges they face in raising funds, the role of state support, legitimate business and criminal activities. It is to a more general understanding of the key characteristics of this typology to which we now turn.

The Goals of Terrorist Financing: Maintenance and Operational Requirements

While al-Qaeda was routed in Afghanistan in 2001–2, in 2016 the multilateral anti-money laundering and counter-terrorist financing organisation, the Financial Action Task Force (FATF), was still claiming that terrorist threats had 'globally intensified considerably'.[31] The rise of the Islamic State in Iraq and Syria (ISIS) contributed to that analysis but a range of international terror attacks by various groups, including al-Qaeda offshoots, terrorist groups in Syria and a resurgent Taliban, also led to growing concerns that terrorist financing was becoming an increasingly significant problem and that there was a continued need to gain a better understanding of the related financing typology.

In comprehending the general terrorist finance typology, the first issue that needs to be addressed is why terrorists need money and how these needs differ from those of criminal money launderers.

The requirements of terrorists for financing specifically fall under two general heads: those relating to maintenance and those relating to operations. Of these two categories, the first, maintaining and expanding organisational structures, is the most expensive. By organisational maintenance we are referring (not exclusively) to the expenses required to run the organisation which supports and directs the operational activity. Expenses here would be all those not directly involved in particular operations. They can include, for example, expenses associated with hiring or buying land for a base and for training, the training itself, weaponry and a multiplicity of assets such as computers and trucks, salaries and associated costs such as death-in-service benefits, facility support, recruitment, propaganda and social services. When terrorist groups control territory and significant populations, then the range of expenses increase exponentially and may equate to some of the spending requirements of a legitimate state.

Operational expenses, by which we mean those directly involved in specific planned attacks, would include funds involved in travel to and from the attack site, expenses associated with surveillance, the use of vehicles, living expenses, and the weapons and components to be employed in the attack.

One of the defining characteristics of terrorist operations is the relatively small cost involved in the actual terrorist attacks. In this regard the 9/11 cost of nearly $500,000 is not reflective of the general costs of terrorist operations, which rarely exceed more than a few tens of thousands of dollars and often are much less.

A Norwegian security services report on small-cell terrorism estimates that approximately 75 per cent of forty violent terrorist plots in Europe between 1994 and 2013 cost less than $10,000. As to lone wolf attacks involving a single terrorist acting independently of any central guiding authority, it is the weapons used in the operation that are the most expensive items, and if they are cars or knives then the costs can be considered truly negligible.[32] For larger terrorist groups such as Boko Haram and al-Qaeda in the Maghreb, the cost of maintenance dwarfs that of operations. But for the small group and lone wolf terrorists, both maintenance and operational costs are very limited indeed.

Table 2.1: Estimated Costs of Terrorist Operations

Terrorist Attack	Date	Operational Cost (est.)
Bishopsgate Bombing	1993	$5,000
East African Embassy Bombings	1998	$50,000
USS *Cole* Attack	2000	$10,000
September 11	2001	<$500,000
Djerba Bombing	2002	$20,000
Bali Bombing	2002	$50,000
Jakarta JW Marriott Hotel Bombing	2003	$30,000
Istanbul Truck Bombing Attack	2003	$40,000
Madrid Transport Bombing	2004	$10,000
London Transport Bombing	2005	$12,000
Printer Cartridges in Air Freight	2010	$4,200
Boston Marathon Bombing	2013	<$500
Kenyan Mall Attack	2013	$2,000
Tunisia Beach Attack	2015	$1,000
Charlie Hebdo Attacks	2015	<$10,000
Nice Truck Attack	2016	<$300
Berlin Xmas Market Attack	2016	<$300
London Westminster Bridge Attack	2017	<$300

The data in this table are derived from several sources including FATF, *Terrorist Financing*, 2008.

Terrorist Financing: 'Money Laundering on Its Head'

As the al-Qaeda example clearly demonstrates, the generation of funds is not a primary objective of terrorists, as it is for most criminals, though terrorists too have to engage in generation and transport processes in order to support organisational maintenance and operations. While the ultimate objective of criminal funds is to support the lifestyle of the criminal, that of the terrorist is to finance terrorist actions or to maintain and grow their organisations.

In order to acquire their desired lifestyle, criminals launder their funds to break the link to the underlying predicate crime. By definition the starting point is the criminal activity while the end point is a patina of legitimacy. But for a terrorist the process may be inverted. That is, the original activity generating the funds may be totally legal. It is only the end to which the finances are put that is illicit. That is why terrorist financing is sometimes referred to as 'money laundering on its head'. The direction of action from illegitimate to legitimate may well be reversed in this case.

Of course the criminal may engage in legitimate business but that, as in the case of Al Capone's laundromats—is often a means of placing funds, or a consequence of integration activities and the final stage of their journey as they try to move away from their roots into the safer legitimate sphere. For terrorists, legitimate business may be a key element of their fund generation methodology but it is not for them a means of pursuing the good life (at least in this world). Rather it may serve as a necessary stepping stone on the way to terrorist action.

So, as we noted above, while in Sudan al-Qaeda financed itself with a range of legal activities including construction companies, tanneries, import-export firms and finance companies, the goal of these activities was to support the organisation's global terrorist campaigns. Groups active in West Africa have sourced money from used car dealerships and the import of motor vehicles to finance their regional activities.[33] In Europe, terrorists have relied on legal pursuits such as salaried employment, social welfare payments, personal loans taken out never to be repaid, and support from family and friends to undertake terrorist activities or to move to another country to engage in similar actions. FATF has reported that the reliance by terrorists on legitimate businesses is especially a problem in areas of the economy where there is no requirement for formal qualifications and also where starting a business is not dependent upon the investment of large amounts of funds. The risk of businesses diverting funds to support terrorism is higher in cash-intensive enterprises where the relationship between sales reported and actual sales is difficult to verify.

Most significantly, however, legitimate businesses and other legal activities for sourcing funds create problems of identification for counter-terrorist financing. Bear in mind that most of the millions and millions of financial activities that occur on a daily basis are entirely legal. If the authorities are seeking to identify an illegal business, they can place the related illicit actions against a backdrop of legal transactions which should help highlight the illegal activities' unique presence. If the activities the authorities are trying to identify are legal themselves, then it becomes even more difficult to identify their existence as they fit readily within the broader context and do not stand out in stark relief. And this is the crux of the problem. Rather than having simply to search for needles in haystacks, counter-terrorist financing needs to be able to identify particular needles within a haystack consisting of needles.

The Terrorist–Criminal Fund Generation Nexus

While the move from legitimate funding to illegal action is one of the key defining elements of the terrorist financing typology, the direction from legitimate to illegitimate activity is not always reflective of what happens on the ground. Indeed, groups including Hamas, Hezbollah and the Revolutionary Armed Forces of Colombia (FARC) demonstrate a hybrid crime–terror modus operandi. The key difference compared with criminals is that the funds resulting from the crime are not geared towards placement and then lifestyle enhancement but are rather to be used for prosecuting terrorist activities.

Table 2.2: Terrorist–Criminal Nexus Examples

Criminal Activity	Terrorist Groups
Drug trafficking	Shining Path, IRA, al-Qaeda, Revolutionary Armed Forces of Colombia (FARC), Taliban, Boko Haram, Hezbollah, AQAP
Extortion	Abu Sayyaf group, Euskadi Ta Askatasuna (ETA), IRA, al-Shabaab, ISIS, Tehrik-i-Taliban (Pakistan)
Kidnapping	Hezbollah, Lashkar-e-Jhangvi, National Liberation Army (ELN, Colombia), Lord's Resistance Army (LRA, Uganda), Hamas, Islamic Army of Aden, Loyalist Volunteer Force, Harkat-ul-Jihad al-Islami (HUJI, Pakistan), al-Itihaad al-Islami (Somalia), ADF (Uganda), Al Qaeda in the Islamic Maghreb (AQIM), Al Qaeda in the Arabian Peninsula (AQAP), Tehrik-i-Taliban (Pakistan), Abu Sayyaf group, Boko Haram, al-Shabaab, ISIS, FARC
Robbery	Abu Sayyaf group, ETA, 17 November (Greece), IRA, Revolutionary People's Liberation Army (Turkey), ISIS
Fraud	IRA, al-Qaeda
Smuggling	IRA, Shining Path, FARC, United Self-Defence Forces (Colombia), ISIS, Kurdistan Workers' Party (PKK), Real Irish Republican Army (RIRA), Hezbollah, Hamas, Boko Haram, ISIS, Lord's Resistance Army
Piracy	al-Shabaab, AQAP

The data in this table is derived from several sources.

Terrorists may thus, alongside their essentially legal activities, engage also in petty crimes in order to fund their operations, though for the larger terrorist groups their criminal activity may extend to more serious and organised crimes. Sometimes the terrorists are also criminals, and so it is natural for them to engage in such activity. And, in any event, terrorists, like criminals, are living and working in an illegal space, so it is not unexpected for them to exploit similar networks and measures. While legal incomes provide security for terrorists because such activities are unlikely in themselves to attract attention, they can require various documents, creating paper trails that can be followed by the authorities. In addition, the yields of legal financial activity are generally not as high as those from criminal financial activities; this, however, may not be a problem because the quantities of monies required may be low. On the other hand, such fund generation may not be as rapid as with illegal activities, and this too may be problematic for the terrorist.

Common criminal methods used by terrorists include drug trafficking, kidnapping for ransom, smuggling, extortion and fraud.

Drug Trafficking

FATF has provided a theoretical model of how a terrorist group may launder drug-generated funds, highlighting the key patterns of the process. So, for example, a terrorist group produces or oversees the production and sale of drugs. A confederate manages a legitimate import-export business in the same jurisdiction where the drugs are sold. The business serves as a front company and a cover aimed at disguising the main source of income, which is the sale of drugs, by mixing that trade with trade-based legal transactions. The confederate manager transfers the money from the jurisdiction of sale to an international financial centre (IFC) where large amounts of global financial activities occur daily. From the IFC some of the funds are then transferred to an off-shore company (presumably where there is limited transparency with regard to beneficial ownership) and from there to various accounts in third countries (or, alternatively, directly to the terrorists) where they are integrated into the legitimate economy or used by the terrorists for various purposes.[34]

There are of course no shortages of actual examples of terrorists engaged in drug trafficking, which remains one of the most important sources of finance for terrorists. In South America, for example, there has emerged a degree of overlap between terrorist and criminal gangs especially in relation to the exploitation of drug resources. The Revolutionary Armed Forces of Colombia (FARC) has in the past relied heavily on the trade with the result that conflict and cooperation were both prevalent among traditional criminal producers and smugglers and the political movement. In the Middle East, the Kurdistan Workers' Party (PKK) is reportedly involved in heroin trafficking along the Turkey–Iran border and sells into Central and Western Europe. According to Turkish police sources, who obviously are in conflict with the Kurdish group, the PKK collects taxes on each kilogram of heroin trafficked to Turkey from Iran, with profits reaching US$200 million annually. Investigations revealed that the PKK organisation played a substantial role in the cultivation, production, trafficking and street-level distribution of various drugs as well as imposing taxes in Turkey and European countries on the traffickers' incomes. Hezbollah, too, is involved in the drugs trade, as are other terrorist organisations.[35]

More recently the focus of the criminal–drug trading nexus has turned to South Asia where at one stage over half of the Afghan Taliban's senior leadership was listed in United Nations Security Council Resolution 1988 as being involved in drug trafficking. Indeed, the UN al-Qaeda sanctions monitoring team claims that for the 2011/12 period approximately a third of the Taliban's annual budget of $400 million derived from the poppy trade. The monitoring team outlines the many elements of fund generation linked to this industry. Poppy farmers pay the Taliban a 10 per cent tax on the opium; Taliban fighters help earn their keep by working in the opium harvest; small traders who collect the opium paste pay the Taliban a tax; and truckers pay a transit tariff. The Taliban is also paid a fee for protecting the opium laboratories. But the biggest source of finance for the Taliban is the payment made by the drug trafficking organisations to the Quetta Shura, the Pakistan-based Taliban leadership.[36]

Taxes are only one part of the financial underpinnings in respect of the Taliban drug trade. The other involves the movement of drugs and the laundering of drug-derived funds. The Taliban example mirrors the theoretical model provided by FATF and described above. The proceeds of drug trafficking accumulate in the countries where the drugs are consumed, mainly in the form of cash. This is a situation not dissimilar to the Black Market Peso Exchange scheme used by drug cartels in South and Central America. Note that all three stages of drug proceeds—laundering, placement, layering and integration—may take place in the consumer country. The findings of the FATF Taliban study suggest that the majority of drug proceeds generated in the consumer countries are transferred to IFCs such as Dubai in the United Arab Emirates (arguably the most important centre here) but they also go to the United Kingdom, China, Democratic Republic of Congo, Iran, Kenya, Lebanon, Malaysia and Pakistan. Integration and placement may take place by way of investment in real estate, securities and shares. Money may also be forwarded to the drug-producing and trafficking countries. Alternatively, drug proceeds may be laundered under export contracts to supply goods to producing and trafficking countries by way of trade-based money-laundering schemes.

Kidnapping for Ransom (KFR)

According to US government sources, between 2005 and 2012 terrorists raised approximately $120 million as a result of kidnapping for ransom

(KFR), while in 2014 alone ISIS generated between $20 and $45 million in ransom payments. Groups employing KFR often exist in ungovernable spaces where the ability of authorities to combat this type of operation is weak. As a result, KFR is prevalent in the African Sahel, Afghanistan and Iraq. Al-Qaeda in the Maghreb (AQIM), al-Qaeda in the Arabian Peninsula (AQAP), ISIS, FARC, Boko Haram and various Pakistani groups have all engaged in KFR. For groups such as AQIM, kidnapping remains one of the main sources of group funding.[37]

Kidnapping—the confinement without legal authority of a person, against his or her will—has of course long been a criminal activity but the object is usually financial, the ransom sum being the end itself. For terrorists, however, the goals can be both financial (a means of raising much-needed funds) and political (to undermine and put pressure on local and overseas governments as well as to win various political and military concessions). Criminal ransom demands are usually lower than those of terrorists, and locals are often targeted by criminals rather than Western nationals, who are the preferred target of many terrorist groups. Targeting foreigners often comes with increased revenue but also increased risk as both local and foreign governmental authorities may more likely become involved. Research certainly indicates that terrorist groups engaged in KFR profile their victims, who are often citizens of Western countries whose governments have a record of paying ransoms,[38] thus tending to support the argument that paying ransom monies to terrorists only encourages future kidnapping.

Some sources are explicit that AQAP employed ransom money it received for releasing returning European hostages to fund the seizure of territory in Yemen in 2011 and 2012. AQIM received $30 million in ransom in 2013 for the release of four French hostages, while in 2013 Boko Haram obtained millions of dollars for the release of eight French citizens kidnapped in northern Cameroon. Al-Shabaab may have received $5 million in exchange for two Spanish hostages kidnapped in Kenya in October 2011.

KFR may be linked not only to fund generation but also directly to political concessions. Qatar reportedly paid a very large sum of money to release members of its royal family kidnapped in Iraq on a hunting trip in late 2015. Reports indicate that between $700 million and $1 billion dollars may have been paid by the Qataris to Shi'ite groups in Iraq, Syria and Lebanon. The incident began when an Iranian-backed Iraqi Shia militia, Kata'ib Hezbollah, an organisation that has ties to Hezbollah in Lebanon, kidnapped the Qataris. The hostage transaction was reportedly linked not only to the large financial payment but to a separate political agreement known as the '4 Towns Deal', which involved the evacuation and ethnic cleansing of four towns in Syria. Two predominantly Syrian Sunni towns, Madhya and Zabadani, which are located near the Lebanese border, were evacuated of their Sunni residents in exchange for the evacuation of Shi'ite residents from the towns of Fua and Kefraya, located in the north-west of Syria. The result of this exchange was to strengthen Hezbollah's position in that country by removing Sunni strongholds near predominantly Shi'ite areas while at the same time protecting Shi'ite civilians who had been besieged by jihadi groups.[39]

One of the attractions of KFR is that kidnappings can occur in one jurisdiction but the ransom may be paid in another. Moreover, various actors may become involved in the ransom payments (there is a veritable industry associated with negotiating KFR). Families and related parties may deal directly with the kidnappers or through governmental authorities. Victims' families may not always choose to notify the authorities that they are negotiating with terrorist groups as this may escalate the problem politically, and political considerations may come to predominate. Payments if made can be deposited in special bank accounts designated by the kidnappers or may pass through alternative remittance systems.

The consensus has thus emerged that 'KFR remains one of the most frequent and profitable source (sic) of illicit financing, and an extremely challenging [terrorist financing] threat to combat'.[40]

Other Criminal Activities

Terrorists engage in smuggling in order to generate funds, and to avoid taxes, controls, embargoes and sanctions. Tobacco has emerged as a useful smuggling item especially in West Africa. Al-Shabaab, in turn, has been involved in ivory trade smuggling, which resulted in no less a figure than Hillary Clinton identifying the process as part of international terrorism. And of course there is also the example of the smuggling of antiquities out of Iraq—a subject we will consider when we review ISIS financing. Examples of armed robberies include the case of the 2002 Bali attack undertaken by Jamal Islamiyah in Indonesia, which was purportedly financed by way of a bank robbery. In terms of extortion (what is sometimes referred to as 'pseudo-sovereignty-based funding'),[41] the Taliban and the PKK use this method to force local populations under their control to contribute funds to the organisation, while groups such as the Liberation Tigers of Tamil Eelam have reportedly engaged in this practice against diaspora communities in Canada. Fraud in all its manifestations, including tax fraud, insurance fraud and credit card fraud, appears prevalent among European jihadi terrorists. Human smuggling and arms trafficking are further fund-generating criminal activities engaged in by terrorist groups. FATF has also noted the emergence of human trafficking as a money-laundering predicate offence and as a potential source of funding for terrorist groups.[42]

In the United Kingdom, where there are no large-scale fundraising efforts for terrorist groups, terrorist financing activity is characterised by the generation of small amounts of money sent to persons overseas for the purpose of funding operations in foreign territories. Methods used have included self-funding fraud such as online fraud and the abuse of student loans and state benefits. Finance for attacks conducted during 2017, for example, included stealing or hiring vehicles, which for lone wolf attackers turned out to be the most significant expense.[43]

Criminal actions are believed to represent an increasingly significant financing source for international terrorism, and according to various studies they represent the largest funding source for European jihadi terrorism. For example, several radical Islamic preachers active in Europe, including the now deceased Anwar al-Awlaki and the previously London-based Abu Hamza, supported criminal actions targeting non-Muslims as a means to finance jihad terrorism.

According to the US Congressional Research Service study on terrorism and criminal linkages: 'Whereas in previous decades criminal and terrorism links occasionally occurred, such connections appear to be taking place with greater frequency today and may be evolving into more of a matter of practice rather than convenience.'[44] Decreased state support for terrorism and tighter regulatory and legal frameworks may strengthen this tendency. Terrorist groups may then move along a continuum of crime and terrorism,[45] passing to and from legitimate to illegitimate fund generation activity as it becomes transactionally attractive or necessary.

Of course, criminal activities come with dangers for the terrorist group, including being caught by the authorities while engaged in what are essentially activities peripheral to their main goal. In addition, engagement in criminal enterprises may engender corruption among politically directed groups leading ultimately to ideological dilution. This phenomenon of ideological dilution can clearly be seen in the case of Northern Ireland. Here dissident Republican groups have increasingly become involved in organised crime, though these crimes are not in themselves specifically related to generating funds for terrorist activities. Fundraising methods include extortion, robbery, cigarette smuggling, and benefit fraud as well as the use of charities and involvement in legitimate businesses. According to the UK *National Risk Assessment of Money Laundering and Terrorist Financing 2017*, 'the lines between raising finance for [dissident Republican] groups and personal gain are ... often blurred'.[46]

Distinctive Sources of Terrorist Finance

Criminally sourced finances are only part of the terrorist funding typology and in some cases not even the most important one. There are also sources unique to terrorists, some of which are generally not available to criminals. These include state support and non-profit organisations; in the case of lone wolf terrorists, personal funding; and in the instance of groups controlling territories and the levers of state, control of natural resources and access to a tax base.

State Sponsorship

One of the key differences between criminal groups and terrorists in terms of financing is that states may directly support terrorist organisations. Criminal groups do not usually receive such support though of course various modi operandi may be reached between states—or elements of the state—and such groups. The symbiotic relationship between some South American states and drug lords or terrorist groups such as FARC in Colombia and the Shining Path in Peru are examples of this latter phenomenon, as is the relationship between some West African states and terrorist groups and criminals.[47]

When we refer to financial support by a state for terrorist groups we are not only referring to the direct supply of financial resources, but the provision by a state of its financial structures and systems so as to supply, transfer and launder funds to and for the terrorists. The state here plays a key role in the placement, layering and integration of these funds. The central implication of this type of relationship is that it makes the actual identification and control of such funds extremely challenging as the state has the ability to create significant opacity.

Why do states support terrorist groups? They do so for a range of military, political, ideological and security reasons. Pakistani support of Kashmiri terrorist organisations, Arab state support of certain Palestinian groups, and the Arabian Gulf state support of various groups in Syria are but a few examples of the exploitation by states of terrorist organisations to support their foreign and security policies. As these examples tend to highlight, state sponsorship of terrorism—including financial support—touches directly upon that vexed definitional problem of what constitutes a terrorist and the trite but inescapably true point that often one man's terrorist is another's freedom fighter. This is naturally an extremely sensitive political issue.

While state sponsorship represents one of the key means by which terrorist groups receive finance, FATF, despite having produced typology studies of everything from football to tobacco, does not produce a typology of state sponsorship of terrorism. This is because no consensus can easily be reached on this vexed problem. FATF does emphasise that state sponsorship of terrorism is incompatible with adherence to FATF standards and principles (which shall be discussed in Chapter 4) and with the UN International Convention for the Suppression of the Financing of Terrorism and other relevant efforts to combat terrorist financing.

Since 1979, the State Department has kept a list of countries that are alleged to have 'repeatedly provided support for acts of international terrorism'.[48] If so designated, a state is subject to a range of unilateral sanctions, including a ban on arms-related exports and sales, prohibitions on economic assistance, and various punitive steps. When first released in 1979, the list included four states: Libya, Iraq, South Yemen and Syria. Only one of those states—Syria—remains on the 2018 list.

For the US Secretary of State to designate a country as a state sponsor of terrorism, he or she must determine that the government of that country has repeatedly provided support for acts of international terrorism. Then, once a country is so designated, it remains so until the designation is rescinded in accordance with certain statutory criteria. In 2018 there were four countries designated under these authorities: Iran, Sudan, Syria and North Korea.

Table 2.3: State Sponsors of Terrorism: US State Department Designations (as of 2018)

Country	Designation Date
Iran	January 1984
Sudan	August 1993
Syria	December 1979
North Korea (re-designation)	November 2017

Iran

Iran was designated a state sponsor of terrorism in 1984. This country supports Hezbollah, which receives direct financial support estimated at hundreds of millions of dollars, and also Hamas though support for it has waxed and waned. The United States also designates as terrorist groups Shia militias including Kata'ib Hezbollah, which Iran has used to fight ISIS as well as spearhead its power in Iraq. According to the State Department, the Iranian Revolutionary Guard Corps—al-Quds Force (IRGC-QF) is Iran's primary mechanism for supporting terrorists abroad. Iran has also provided weapons and finance to Shia militias in Bahrain as well as to Palestine Islamic Jihad and the Popular Front for the Liberation of Palestine (PFLP) General Command. According to the State Department, Iran permits al-Qaeda operatives to function in Iran and has allowed them to use Iran as a financial hub. In the 2015 US Department of State *Country Report on Terrorism* Iran was identified as the world's main sponsor of terrorism.[49] The 2016 report indicated little improvement in this area on the part of Teheran.[50] In 2019 the IRGC-QF was designated by the US as a terrorist organisation.

Sudan

Sudan was designated as a state sponsor of terrorism in 1993 because of its support for the Abu Nidal Organisation, Palestine Islamic Jihad, Hamas and Hezbollah. In the mid-1990s Sudan served as a centre, safe haven and training hub for various terrorist groups including al-Qaeda (for which it also acted as a financial hub). Hamas has, according to the State Department, also raised funds in Sudan.

Syria

Syria was designated in 1979. According to the State Department's *Country Report on Terrorism* in 2015, the Assad regime's economy raised great concerns as it was extremely vulnerable to exploitation for terrorist financing purposes. In Syria, 60 per cent of all business transactions are conducted in cash and nearly 80 per cent of Syrians do not use formal banking services. Despite Syrian legislation that required money changers to be licensed, many continued to operate illegally in Syria's vast black market, estimated to be as large as Syria's formal economy. Regional informal banking networks

remained heavily involved with smuggling and trade-based money laundering, activities which continue to be facilitated by corrupt officials. The
Syrian state has in the past provided financial and other support to
Hezbollah in exchange for military support and has allowed Iran to use
Syria as a base to support Hezbollah, though much of this financial support
for the Lebanese Shia organisation has been replaced by Iran. Syria has,
according to the State Department, displayed a permissive attitude towards
al-Qaeda and other groups fighting in Iraq. In respect of al-Qaeda it has
acted as a transit point, financial and otherwise. It should be noted that the
Syrian government continues actively to fight al-Qaeda and associated
forces in various parts of the country.

North Korea

The country was designated a state sponsor of terrorism in 1988 and stayed on
the list until it was removed in 2008. North Korea had for several years been
implicated in various terrorist attacks, including the 1983 attack in Rangoon
that attempted to assassinate the President of South Korea, Chun Doo-hwan,
and which succeeded in killing 21 persons. The immediate cause of the North
Korean designation was its bombing of Korean Air flight 858 in 1987, which
killed all 115 people on board. A key reason for removing North Korea from
the list in 2008 was the hope by the Bush administration that such a step
would help ongoing nuclear disarmament talks. The State Department's 2007
Country Report on Terrorism thus stated that North Korea was not known to
have sponsored any terrorist acts since 1987 and the US government noted
that Pyongyang had issued 'an authoritative and direct public statement
affirming that it does not support international terrorism now and will not
support international terrorism in the future'.[51] The re-designation of North
Korea as a state sponsor of terrorism followed its hacking of Sony Pictures in
2014 and its murder by chemical weapons of Kim Jong-un's half-brother, Kim
Jong-nam. More significantly, the re-designation was part of the effort by the
Trump administration to ratchet up pressure on North Korea in response to
its continued nuclear and missile testing.

Removing a State from the List

Technically, for a state sponsor of terrorism to be removed from the State
Department list, the President must report to Congress and certify that there

has been a fundamental change in the leadership and policies of the government of the country in question, that the government is not supporting acts of international terrorism and, further, that the targeted government has provided assurances that it will not in the future support acts of international terrorism. Alternatively, the President may submit a report to Congress justifying the proposed rescission and certifying that the government concerned has not provided any support for international terrorism during the preceding six-month period and that it has provided assurances that it will not do so in the future.

In addition to the North Korean example, Cuba, Iraq, Libya and the former South Yemen have in the past been removed from the list. Arguably their removal was due to broader considerations: Libya and North Korea for non-proliferation reasons, and Cuba for more general foreign policy considerations. The fact is that counter-terrorist financing policy is often subordinated to other, more pressing security issues.

Interestingly, various countries that have been widely accused of supporting terrorism do not appear on the list, among them Saudi Arabia and Pakistan (though both deny such accusations). Countries such as Yemen, Qatar and Eritrea have in the past been accused of financially supporting al-Shabaab, a charge which they all deny.[52] While it could be argued that these states do not support terrorism and should not therefore appear on the list, one could also say that broader security considerations trump counter-terrorist and terrorist funding concerns.

The Role of Charities

In the United Kingdom, for example, there are over 900,000 organisations in the non-profit sector but the vast majority of these are now recognised to be at low risk of terrorist abuse. Rather, the risks are located in a sub-sector of under 20,000 charities which tend to be newly registered, have low income, and operate globally, especially in high risk areas such as Syria and Iraq.[53]

While criminals do not readily receive charitable contributions, terrorists do—for political and ideological reasons. Private donations from individuals in the Arabian Gulf have historically been very important not only for al-Qaeda, but for various Sunni groups in Syria, Iraq and elsewhere. Individual donors in Kuwait and Qatar as well as Saudi Arabia, Pakistan and Turkey have been crucial here. As noted by one US Undersecretary of State working in the field of financial intelligence: 'private fund raising networks rely increasingly upon social media to solicit donations and communicate with donors and recipient opposition groups or terrorist organisations'.[54] So charities are undoubtedly an important source of funding. Certainly religious charitable (*zakat*) contributions have for many years been subject to a light regulatory regime within the Arabian Gulf. Charities in that region have raised hundreds of millions of dollars through fundraising in mosques and homes and on social media. Arabian Gulf states too have channelled money into a range of Wahhabi and Salafi charities, some of which may have or had contacts with terrorist groups. An example here would be the al-Haramain Islamic Foundation, which had ties to the Saudi Arabian government and which according to the FBI provided financial and logistical support to al-Qaeda.[55] The US Treasury Department has specifically identified Qatar and Kuwait as permissive environments for donor-driven terrorist financing.

Of course, private donations can morph into extortion though not always. According to Europol reports, both the Taliban and the Kurdistan Workers' Party (PKK) extort money not only from captive populations but also from diaspora communities. For example, according to Human Rights Watch, the Liberation Tigers of Tamil Eelam (LTTE) extorted 'charitable' monies from the Tamil communities in Canada. Al-Shabaab extorted 'charitable' funds from the Somali diaspora, in the United States, Europe and elsewhere. Hezbollah receives donations from Lebanese Shia communities in Europe, the Americas, Africa and Asia, and Hamas does likewise from Palestinian communities throughout the world, collecting them through charitable organisations such as the Holy Land Foundation in the United States. Lashkar-e-Taiba receives its funds from Pakistan, including by way of private donations and by using front organisations such as Jamat-ud-Dawa and its charity wing, the Falah-e-Insaniat Foundation, to secure funds.[56] The line between willing and coercive donations is sometimes blurred.

Whatever the case, certainly prior to 9/11 charities and NGOs received relatively light government oversight. The nature of their 'good works' often meant that such institutions benefited from a benevolent attitude to their functioning. But this lack of oversight made such organisations attractive to terrorists. The links to good works attracted unwitting donors and the geographical dispersal of charitable branches meant that money could be raised in one jurisdiction, where there were contributors with funds, and easily transferred to another, including jurisdictions where terrorist cells and organisations were located or attacks planned.

Research has identified five key methods by which terrorists exploit non-profit organisations and charities, though these are not always clearly distinguishable.[57]

- Diversion. This involves the diversion of donations by affiliated persons functioning within the charity or NGO to terrorist organisations. In this most common form of charitable vulnerability, funds raised for humanitarian programmes are diverted to support terrorism at some point during the charity's business cycle. The penetration takes place during all three stages of a charity's work—collection, retention and transfer. In the collection phase the money is intercepted before it is deposited into the charity's accounts. During the latter two stages, the funds are diverted from the public objectives of the charity to the intended terrorist objective. A distinction can also be made here between those carrying out the diversion who are internal to the group and those external to the charity. One example of the latter is that of Muslim Aid in Birmingham, UK, where certain individuals fraudulently presented themselves as collecting money for a legitimate charity and then diverted funds that had been collected by that organisation to fund terrorist activities.[58]

- Affiliation. Here the charity is affiliated to a terrorist organisation. Specifically, a charity or an individual working for a charity maintains an actual operational connection with a terrorist group. The nature of the affiliation may range from formal to informal. Compared to the diversion model, there is here a much tighter connection between the charity and the terrorist organisation; the former effectively functions as part of the latter. An example often mentioned in this context is the Melbourne-based Tamil Coordinating Committee, which collected money from the Tamil community in Australia for the Liberation Tigers of Tamil Eelam. The group was run by individuals directly involved in terrorist activities against the Sri Lankan government.

- Abuse of programme delivery. Charitably funded programmes can be diverted at the point of delivery to support terrorism. One example of this was where a charity set up to promote education and religion complemented its charitable goals with an online list of suicide bomber 'martyrs' published by a terrorist sympathiser who had infiltrated the charity.[59] Charities supporting 'service activities' operating close to battle zones tend to be particularly vulnerable to programme delivery abuse. The Australian government has, for example, voiced concerns that some funds sent to Syria and neighbouring countries are vulnerable to terrorist exploitation if they are sent through start-up charities that do not have proper due diligence procedures in place.

- Support for recruitment of terrorist organisations. Here the charity or NGO uses its facilities or programmes to support recruitment of personnel for terrorist organisations. Charities can also supply funds to support the families of terrorists—a very attractive recruitment incentive—or employ their facilities to train and recruit terrorists as well as hosting speakers supporting terrorist organisations. FATF refers to an example of the Pakistani-based al-Rehmat Trust, which provided financial support to designated terrorist organisations including al-Qaeda.[60]

- Sham charities. This involves the creation of sham charities or NGOs through false representation. Here the charity is really a front for the terrorist group and the stated purposes of the charity are totally false.

FATF in its 2014 study has estimated the frequency of abuse of these methodologies, with diversion of funds clearly being the most prevalent. As multiple methods were often used, the figures in Table 2.4 do not add up to 100 per cent.

Table 2.4: Terrorist Abuse of Charities and NGOs

Methods and Risk of Abuse	Frequency Observed
Diversion of Funds	54%
Affiliation with a Terrorist Entity	45%
Abuse of Programming	10%
Support for Recruitment	26%
False Representation	14%

Taken from FATF, 'Risk of Terrorist Abuse in Non-Profit Organisations', p. 37.

During an investigation into the transfer of funds from the United Kingdom to foreign terrorist fighters in Syria, the Charity Commission of England and Wales uncovered an organisation involved in the solicitation of funds for aid convoys to the war zone. These aid convoys were recognised as increasingly vulnerable to various methodologies of terrorist finance abuse as the aid was intended to be diverted to terrorist fighters. Intervention by the authorities led to the winding-up of the organisation and ultimately a reduction of these aid convoys.[61]

Greater regulatory oversight has complicated but certainly not eliminated this financing method and it remains a critical source of funding for many terrorist groups. There is also another variant where the terrorist group uses charities as an arm of the state or to fill gaps left by weak or failed states. We will look at these when we come to terrorist groups that control territory and the levers of statehood.

Self-Financing

That self-financing represents a distinct source of terrorist financing is a reflection of the fact that terrorist operations are relatively cheap to finance and that, unlike criminal financing and laundering, the money is not an end in itself. With the increasing emergence of small cell and lone wolf attacks, self-funding as a money-generating process has become increasingly important.

In a Norwegian study of jihadi cells in Europe, it was clearly demonstrated that the majority of the 40 cells studied relied on more than one source of finance but cell members' personal assets were by far the most common funding source. According to the data reviewed, 58 per cent of the cells relied at least partly on the cell members' salaries, welfare payments and savings to finance their activities. The most common income sources of European terrorist cells were salaries and savings of cell members, followed by trade in drugs, cars, forged documents, weapons and other goods, and finally theft and robbery. Significantly, the majority of jihadi cells studied received no financial support from global terrorist organisations.[62]

That small cells and lone wolf attacks are self-financed may be readily explained by the fact that they need less money for maintenance or operations. As noted in the Norwegian study, single-actor terrorists are overrepresented in the category of self-financing and several of their attacks were perpetrated with the use of hand-held weapons of negligible cost. Nevertheless, there are exceptions to this phenomenon. The shoe bomber Richard Reid and the 'underwear bomber' Umar Farouk Abdulmutallab received external support while some larger cells such as the 'Doctor Cell', which attacked Glasgow Airport, and that responsible for the London Underground bombings in 2005 were self-financed.[63] The Charlie Hebdo and Kosher Store attack in Paris in 2015 were funded by a €6,000 loan obtained with forged documents, the proceeds of the overseas sale of a used car and cash transfers linked to the sale of counterfeit goods. Significantly, self-financed cells are more likely to launch attacks than other cells. Among the cells in the Norwegian sample that were entirely self-financed, 53 per cent actually went on to execute their plans, compared to 21 per cent of those terrorists that received external support.

In relation specifically to the United Kingdom, FATF had emphasised that the country's terrorist finance risks are often low value, self-funded and derived from legitimate sources.[64] Money service businesses, cash couriers and retail banking are considered the sectors at highest risk of terrorist abuse.[65]

Emerging Trends

In 2015 FATF identified several emerging threats in the terrorist financing space.[66] First there were financial methods associated with social media. Social media and crowd funding certainly represent a means by which larger terrorist groups may raise funds. In 2013 on a Facebook group sharing recipes, a fighter in Syria requested equipment, food and pharmaceuticals. The user provided details of a German bank account to which the funds should be sent.[67] Undoubtedly, terrorists have proved adept at using the internet to communicate, co-ordinate activities, disseminate propaganda, collect intelligence and raise funds. As FATF notes: 'Social networks are widely used by terrorist organisations to spread their terrorist propaganda and reach out globally to sympathisers.'[68]

Secondly, there were the financial considerations associated specifically with so-called foreign terrorist fighters (FTFs) who made their way across Europe to the battlefields of Syria and Iraq in support of ISIS and other terrorist groups active in those countries. Here self-financing has proved to be a very important source of finance.[69] These persons needed financing for food, clothing and transport costs as they made their way from Europe to Turkey and then from Turkey to the battle zones. Studies have indicated that the funds of these FTFs often derived from legitimate sources such as support from families, bank loans, social security payments and earnings from employment. These legitimate sources were complemented by criminal activities such as petty crime and fraud, including bank loans taken out with no intention of being repaid. How FTFs returning from the war zones to their homes in Europe and elsewhere are able to finance their travel is a subject that requires some research.[70]

Most significant was what FATF termed the appropriation of natural resources for profit. This was manifested most strongly where terrorist organisations took control of territory and the levers of state power.

The Control of States and Financing Sources

ISIS was during its 2014 and 2015 heyday the best-financed terrorist organisation in the world and it manifested many of the essentials of financial statehood.[71] Access to a range of economic and financial resources magnified its political and military power. At its peak the group controlled by some estimates more than 90,000 square kilometres of territory, including portions of the Syrian provinces of Aleppo, Raqqa, Deir ez-Zor, Homs, Hasakah and others as well as parts of the Iraqi provinces of Saladin, al-Anbar and Nineveh. According to one estimate, ISIS had an annual revenue in 2014 of $1.9 billion.[72]

There were three key elements of ISIS's financial power stemming from this state and territorial control that were, with perhaps one or two exceptions, sui generis to that organisation. These were (i) control of a not insignificant regional banking network; (ii) control of a wide and potentially lucrative range of natural resources; and (iii) a large tax base in the form of a captive population. While other terrorist groups may have had control of some territory, including natural and human resources capable of exploitation, none matched the scale and scope of that controlled by ISIS.

ISIS's Banking System

In the summer of 2014 ISIS forces took control of Iraq's second largest city, Mosul, and within it several branches of public and private banks, including branches of the Central Bank of Iraq and the headquarters of Mosul Bank. The capture of provinces in Nineveh, Saladin and other areas also secured additional bank branches and, indeed, by 2015 ISIS may have controlled over a hundred of them. Similarly, in Raqqa and surrounding areas they took control of over twenty branches.

Figures in respect of ISIS earnings and spending are notoriously difficult to estimate definitively. According to some estimates, total cash holdings in safes and various accounts may have exceeded over half a billion dollars or even double that amount. No other terrorist organisation has ever matched such funding levels, and these funds proved critical in financing ISIS's expansive military and civilian operations.

Not only did assets located in bank vaults fall into the hands of ISIS, but branches in Iraq continued to function as Baghdad authorities for some time continued to pay salaries to civil servants located within ISIS territory. These payments finally stopped following American remonstrations. The United States also in 2015 began to reduce transfers of dollars to Iraq's Central Bank so as to prevent any access by ISIS to such funds.[73] Moreover, as most people stranded in ISIS territory never had bank accounts to begin with, it is not surprising that the formal banking system was also backed up by an extensive system of alternative remittance systems. These allowed ISIS to trade in, transfer and receive funds with agents in Turkey and Jordan, outside the confines of the formal banking system.

ISIS's Control of Natural Resources

In 2015 ISIS controlled approximately fifteen oilfields in Syria, mainly in the regions of Deir ez-Zor, Rawah and Hasakah, and ten oilfields in Iraq, including the Qayyarah oilfield and the Baiji refinery. Again, while estimates vary considerably, it is believed that ISIS's overall oil production amounted to approximately 80,000 barrels per day (bpd) in 2014. One estimate puts ISIS's earnings from oil production at $435–$550 million in 2015.[74] The largest market for the oil was internal, within ISIS-controlled territory, but a trade of sorts was carried out with the Syrian government in Damascus, complemented by exports employing middlemen and smugglers across the Turkish and Jordanian borders.

Oil revenue was estimated by one report as $1 billion at the end of 2014 and $600 million in 2015. At the time, this provided a source of significant income for ISIS. From then on there followed a precipitous reduction of related income caused by a combination of air strikes as part of the international coalition's Operation Tidal Wave II assault on production, refining and transport systems, and the effects of withdrawal of national and private sector companies capable of operating the oil industry.[75]

Although ISIS gained control of several gas fields (especially after its capture of Palmyra in May 2015), technical demands combined with problems of moving natural gas for export and the effect of air strikes meant that the group could never fully exploit the potential of this natural resource.

ISIS's Tax and Fee Base

At its height ISIS controlled approximately 8 million people. This potential tax base (or object of extortion, depending upon the definition used) became in 2015 one of the primary sources of ISIS finance. That year, ISIS authorities may have generated from various taxes and fees approximately $400–$800 million,[76] about a third of the group's income during this period. Revenue from this stream included taxes on the salaries of employees of the Syrian and Iraqi state who were still receiving payment from those two governments; customs duties on trucks entering ISIS territory; taxes on agricultural products such as cotton, barley and wheat; taxes on economic activities; and a further tax on religious minorities. In addition, ISIS authorities charged various fees for services such as electricity, water and the use of telephones. There were also fees imposed on the rental of public buildings, on bank transactions, the registration of vehicles and educational enrolment.[77] The combination of bank expropriations, resource sales and taxation generated over 90 per cent of ISIS's income during the 2014–16 period.

Other Resources

Part of ISIS's income also derived from a myriad of other sources including kidnapping for ransom, the export of historical and archaeological artefacts, and money brought in by FTFs. There were additional reports of external charitable donations deriving from several Arabian Gulf states including Kuwait and Qatar, but compared to that of transnational terrorist groups, this external source of funding was relatively insignificant.

ISIS Expenses

Unlike the situation for transnational terrorist forces or those whose attachment to the land is more fleeting and tactical, control and deployment of state power circumscribed by a defined territory came with financial responsibilities. While transnational terrorist organisations may have few maintenance or operational expenses, for an organisation that boasted the trappings of a state, the level of associated outgoings was potentially very high.

Most pressingly, funds were employed to support extensive conventional military operations (especially the salaries of fighters, training, recruitment, propaganda and benefits) in several theatres, the group's expansion abroad (particularly in Libya), and the administration of its territories.

Key to ISIS's rationale was the need to present itself as an authentic political entity. Thus it funded an array of social services such as healthcare and schools. Compared to the military, these institutions appear however to have received limited funding. Budget decisions were centralised in Mosul with regional budgets being managed by governors in the various provinces. As the military campaign against ISIS intensified, this infrastructure began to erode rapidly.

The Demise of ISIS: The Financial Debate

ISIS's unique and extensive funding network was recognised early on as a particular problem in that it provided the organisation with enormous resources but it was also identified as a vulnerability, one that required a financial response. In 2015, the Counter ISIL Finance Group (CIFG) was established and co-chaired by the United States, Italy and Saudi Arabia. The group had four key objectives: to prevent ISIS's use of the international financial system; to counter the extortion and exploitation of economic assets and resources that were derived from ISIS territory; to deny ISIS funding from abroad; and to prevent ISIS from providing financial or material support to foreign affiliates in an effort to expand its global ambitions.

These objectives involved, among other things, cutting off ISIS-controlled banks and alternative remittance systems from the global financial system, imposing sanctions on various groups and individuals, disrupting ISIS's oil and gas industry, reducing financial liquidity in ISIS-controlled territory, disrupting the smuggling of antiquities, preventing foreign donations, and calling for non-payment in response to kidnapping for ransom. All of these steps were backed up by financial co-ordination, technical co-operation and intelligence sharing among the international community.

The private sector was brought fully into these efforts and expected to contribute by including anti-ISIS measures within their general anti-money laundering and counter-terrorist financing and sanctions policies. Significantly, of course, most of ISIS's fund-generation capacity was restricted to its own territory and lay beyond the reach of counter-terrorist financing efforts. Nevertheless, the group's interface with the global financial system still represented a vulnerability that needed to be targeted, though even this goal was challenging given that, as FATF noted, ISIS was paid mostly in cash for its oil, 'making the transactions underlying its oil trade difficult to track and disrupt'.[78] The CIFG thus moved to isolate over 90 bank branches located within ISIS territory from the international financial system; helped Iraq's central bank create a list of over 100 money transfer companies and exchange houses working within ISIS territory and then banning them from accessing dollars auctioned by the Iraqi central bank; and as noted, worked (with the key assistance of Washington) with the Iraq government to prevent salary payments to officials within ISIS territory.[79]

From the summer of 2017 ISIS began to lose territory rapidly and the group was later forced out of its bases in Mosul and Raqqa. The question that arises here was: how effective were the specific financial measures adopted, compared to the more traditional military actions, including intensive air attacks and ultimately full-scale ground operations, in hastening ISIS's demise? Arguably, if only financial steps had been undertaken until 2017, then ISIS would still certainly be in the same position as it was at the beginning of 2015. Supporters of the financial campaign would, however, argue that financial intelligence helped identify targets for destruction by the various coalitions targeting ISIS and thus weakened the enemy's power. For example, the Operation Tidal Wave II campaign, which included the bombing of over 25 ISIS cash stockpiles, presumably helped to reduce the ability of ISIS authorities to pay its fighters—though it should be noted estimates varied significantly as to how much cash had actually been destroyed in these operations.[80] Bombing attacks on oil refineries and transport reduced ISIS's economic strength. By the end of 2017 the allies claim to have bombed over 2,600 ISIS energy targets.[81] But whether the financial campaign in itself actually weakened ISIS fighting capability significantly in preparation for the final ground campaign, which removed the organisation from Mosul, Raqqa and other areas, is far from clear. In any event, these bombing attacks, even when targeting economic resources, are more legitimately classed as part of standard military operations rather than part of the financial campaign.

ISIS's financial power was primarily dependent upon its control of territory and its exploitation of resources within that area. As the territory contracted, so did the group's financial power. ISIS's loss of territory was therefore directly a result of the application of military force, not of any financial measures, which had little impact upon either ISIS's order of battle or military prowess. In any event, the global financial campaign tended to target the group's relationship with the global financial system, whereas the key sources of ISIS's financial power were as noted domestic and beyond the reach of global financial forces.

Between 2014 and 2017 the anti-ISIS financial campaign received an inordinate amount of attention, which may have led some to believe that it truly represented an independent military weapon that in itself could have caused major damage and degradation to ISIS military power. Sceptics may have viewed the effort expended on these financial methods as excuses for avoiding direct and more effective military action or the expense involved in upgrading relevant military capabilities. Students of Finance and Security always need to be aware that it is a characteristic of bloodless financial warfare that it is often on the declaratory level confused with real war. For example, in 2017 the European Union's Directorate-General for External Policies stated:

> The European Union is also active in the fight against IS, especially in terms of adopting measures aiming to sever the terrorists' financial networks. The circulation of cash and the transfer of funds are both heavily regulated in order to increase the traceability of financial flows, all the while keeping an eye on digital platforms through which virtual currency can be exchanged.[82]

While this may have made the European Union feel involved and relevant,[83] these kinds of efforts were very subordinate to military actions. In late 2017, as the military campaign on the ground intensified (at the cost of numerous casualties among the anti-ISIS forces and civilians caught in the cross-fire), the report recommended that the European Parliament 'could be well advised to press for a study on the different levels of commitment and progress Member States have realised in the fight against terrorism and, notably, the efforts made to hinder IS' financing'. Further studies on how to counter ISIS financing were fortunately soon overtaken by military action on the ground.

Other Territorial Groups: Hamas and Hezbollah

ISIS is no longer present in its major territorial incarnation, but other terrorist groups have merged into state structures or effectively control the levers of state power and thus are able to use state financial and other resources to enhance their standing.

Hamas took control of Gaza in 2007 and remains subject to an Israeli and Egyptian blockade. Because it controls territory over which it rules, Hamas has at its disposal a tax base. Gaza is also a major recipient of foreign aid from states such as Qatar and Iran and from the United Nations through the activities of the UN Relief and Works Agency for Palestine. A major tunnel system provides a steady source of income though there is debate as to what percentage of Hamas's budget ultimately derives from this network. What is not debatable is that Hamas's handling of the Gaza economy has been disastrous, though, to be fair to the organisation, confrontation with Israel and not economic development has always been its main objective.

Hezbollah presents another model of terrorist control of state resources. Unlike Hamas in Gaza, Hezbollah does not fully control Lebanon but it exerts major influence on the governance of that country. The organisation manifests itself as a powerful—but not all-powerful—political party within Lebanese politics and as a guerrilla insurgency dedicated to fighting Israel and other enemies of the Iranian state, including those within Syria. It has many expenses but does not by itself have sole access to Lebanese state resources or the country's tax base as it functions within a broader coalition. The organisation maintains a powerful guerrilla force especially in the south of the country and also in Syria, as well as a network supporting social and political activities in many Shia areas of Lebanon. These activities serve to fill a gap in services left by weak state organisation while at the same time they enhance

Hezbollah's influence and roots within targeted communities. All these activities require significant resources.

The key source of Hezbollah's funding is foreign state support from Iran. While debate exists as to the size of the financial aid it receives from Iran, there is no question that it is significant and has increased as a result of Hezbollah's involvement in the Syrian civil war. As Matthew Levitt noted in 2005, while Hezbollah had in the past benefited from the usual sources of funds, such as criminal activities, charities and foreign support, 'Unlike most terrorist groups which need to focus much time and attention on raising, laundering and transferring funds, Iran's largesse provides Hezbollah with a sizable and constant flow of reliable funding. By all accounts Hezbollah operates under no revenue constraints.'[84] The size of this support means that despite its key role within the Lebanese state, Hezbollah has limited need for exploiting state resources for maintenance or operational activities.

That being said, there are also reports that Hezbollah has become more deeply involved in international drugs trafficking, weapons smuggling and money laundering (through, for example, the purchase of used American cars which were then exported to Africa) to support its income. Accusations such as these are vehemently denied by the leader of the group, Hassan Nasrallah, who has stated on several occasions that it would not be religiously permissible for members of Hezbollah to be involved in the drugs trade.[85]

By subordinating economic development to military priorities, terrorist groups have been unable to act as effective agents for economic development, but they have been able to exploit the resources of the state to support their more pressing security objectives.

The generation of funds is but one of the challenges that criminal money launderers and terrorist financiers face. The other—for terrorist organisations—is the movement of those funds from the point of generation to the point of attack. This task needs to be undertaken while evading regulatory and security authorities and disguising the links between the funds, the people transferring the funds, and the ends to which those funds are put. It is to this subject that we now turn.

3

THE TRANSFER OF ILLICIT FUNDS

If you were in Venezuela or Ecuador or North Korea or a bunch of parts like that, or if you were a drug dealer, a murderer, stuff like that, you are better off doing it in bitcoin than U.S. dollars.

Jamie Dimon, CEO, JPMorgan Chase[1]

Moving Funds

Criminals obfuscate the illegal provenance of their funds but terrorists aim to mask their murderous objectives. While terrorist finance is thus criminal money laundering on its head, terrorists, similarly to criminals, may still need to move their funds from the point of generation to the point of terrorist target. And in so doing terrorists may similarly to criminals also need to disguise both their own identity and their ultimate intent.

The fund generation process is thus only part of the challenge faced by both the criminals and terrorists. Potentially there are numerous methods by which terrorists can move their funds across borders and from where they are accumulated to where they are needed. Just as with the manner in which the funds are raised, how terrorists move monies bears some continuities with those of criminals. However, there are also some options that are generally not readily available to the criminal but which may present themselves to the terrorist.

This chapter will review the challenges posed to criminals in respect of fund movement and will also focus on terrorists and their transmission strategies.

Specifically, it will analyse the role of cash couriers, the use of the formal banking network, and reliance on alternative remittance systems as well as various key products. Finally, this chapter will assess the opportunities for terrorist and criminal financing presented by the emergence of cryptocurrencies.

The Physical Movement of Cash

If you plan to take significant amounts of cash with you on holiday, on arrival at the airport you best declare these sums to customs. In the United Kingdom you must declare cash of £10,000 or more (or the equivalent in another currency) if you intend to take the monies between the United Kingdom and any non-European Union country (and possibly European Union states as well if you leave after Brexit). Cash here includes notes and coins, bankers' drafts and cheques of any kind, including travellers' cheques. If you don't declare the cash or give incorrect information, you could face a penalty of up to £5,000 and customs officers can seize your monies. The European Union states have similar rules as does the United States, where you must report cash sums in excess of $10,000.[2]

Part of the intention here is to apprehend criminals and terrorists moving money across borders. The carrying of large sums may point to illicit intent though, presumably, neither knowledgeable criminals nor competent terrorists will readily volunteer these declarations if they are indeed carrying such sums. Alternatively, they always have the option to structure their cash below any *de minimis* threshold. Of course, criminals and terrorists may also use such currency declarations as a means of giving legitimacy to their cross-border transfers. United Kingdom authorities for example continue to collect cross border cash reports of which there were 3,747 declarations in 2016 totaling Euro 466,430,404.[3]

Cash remains an important method of global settlement and large amounts of transactions continue to be conducted in this medium; at the same time cash smuggling represents one of the oldest forms of global money laundering. Criminals may engage in cash smuggling for purposes of paying suppliers, banking ill-gotten gains in a foreign jurisdiction or simply breaking the audit trail. The international anti-money laundering and counter-terrorist financing organisation, the Financial Action Task Force (FATF), has noted that the physical movement of cash is employed by criminals involved in a wide range of activities including tax fraud, immigration crime and arms smuggling.[4]

Unsurprisingly, the physical movement of cash also remains among the favourite methods terrorists use to get their funds to where they are needed most. This is no doubt because of the inherent 'anonymity, portability, liquidity and lack of audit trail' associated with the physical movement of cash.[5] Indeed, according to the US Treasury Department, 28 per cent of all terrorism convictions in the United States between 2011 and 2016 involved the physical movement of cash.[6] Between 2007 and 2015, 18 terrorist financing prosecutions in the United States involved the use of cash transfers. These cases involved al-Qaeda, al-Qaeda in Iraq (AQI), al-Qaeda in the Arabian Peninsula (AQAP), al-Shabaab, FARC and Hezbollah.[7]

In the case of Hezbollah, a US citizen was convicted of conspiring to smuggle cash to the Lebanese-based organisation, partly by way of couriers carrying cash structured in amounts of less than $10,000, either directly to Lebanon or via a third country. While this would indicate that the $10,000 threshold at least acted as a complicating factor, during the trial one of the defendants stated that he had in fact carried $66,000 over the course of two trips to Lebanon. There were other plans for concealing $500,000 in a car to be sent to Lebanon on a container ship.[8] There have been other cases involving attempts to place cash in cars and tractors shipped to Lebanon and Iraq from the United States.[9] Certainly freight in the form of passenger vehicles, heavy goods vehicles and commercial airlines remain important means for smuggling cash across borders.

UK authorities have noted that cash couriering is the most favoured method for terrorists taking money out of the United Kingdom and that 'cash is thought to remain the highest risk area for terrorist financing'.[10] Between 1 April 2012 and 31 March 2017, 79 terrorist financing-related cash seizures were made in the United Kingdom, totalling £495,797.[11]

While for the large-scale money launderer, because of the enormous sums that need to be placed, layered and integrated, the physical transport of cash may be secondary to trade-based money laundering, for the terrorist, for whom the sums involved may generally be smaller, cash smuggling may be a key method.

Given the small amount of money needed to launch a terrorist attack, physical movement by cash couriers can be an effective method of transporting funds. The UK authorities provided the example of the withdrawal by an individual between 2014 and 2015 of approximately £3,000 from the bank account of someone suspected of fighting for the Islamic State in Iraq and Syria (ISIS) in Syria and the provision of that cash to a suspect subsequently linked to the Brussels terrorist attack of 2016 for the purpose of forwarding the money to the fighter active in the Syrian war zone.[12]

Another attraction of physical cash movement is that it avoids for the most part the formal banking system and the regulatory oversight integral to that system. We have already noted that foreign terrorist fighters who moved to Iraq and Syria during ISIS's heyday took the limited funds they needed on their person. Following the US invasion of Afghanistan in 2001, both al-Qaeda and the Taliban also physically moved money both across the Pakistan and Afghan borders and between different geographic regions in those countries. The 9/11 report prepared by the US National Commission on Terrorist Attacks upon the United States noted that it was a staple of al-Qaeda strategy to recruit couriers within the organisation to physically transport the cash. The funding for the JW Marriott Hotel bombing in Jakarta in 2003 was provided to the terrorists in the form of cash by al-Qaeda operatives based in Thailand. In addition, a further $30,000 was supplied to Indonesian terrorists by al-Qaeda cash couriers. The al-Qaeda funds for the earlier 2002 Bali bombing reached the local perpetrators also by way of courier transfer.[13]

In any event, in some areas such as parts of Africa or South-east Asia there is little or no formal banking system, and so the movement of cash is the only real means of getting the funds from the point of generation to the point of employment. Boko Haram has made significant use of cash couriers both for moving money into Nigeria from neighbouring countries and within the country itself.[14] FATF has focused on several cases in West Africa where bulk smuggling has been carried out, especially in support of Boko Haram. In 2012 a courier for Boko Haram gave evidence when arrested in north-western Nigeria that women were actually the preferred cash couriers as they were not searched by the security services at roadblocks. Presumably the Nigerian authorities have changed their approach since then. Where male couriers were used they pretended to be commercial drivers moving goods through the region. Other cash couriers for Boko Haram have been detained on the Niger–Nigerian border as well as some emanating from Mali.

The obvious difficulties of detecting relatively small-scale cash movement make this method hard to beat. This no doubt explains both the limitations of this method for criminals who may wish to transfer large sums and its concomitant attractions for terrorists. The US Treasury is convinced that cash smuggling will remain an important means by which terrorists move funds, and it points to al-Qaeda, Hezbollah, al-Shabaab and FARC in this regard.[15] In turn, FATF has noted the increase of bulk cash smuggling across borders.[16] Not surprisingly, both the United States and the United Kingdom deploy specialist cash seizure teams with the aim of disrupting this method of terrorist fund movement.

The Formal Banking System, Correspondent Banking and the Problem of De-risking

Big banks have 'unbanked everyone from porn actors to pawnbrokers', says a regulator.

The Economist, 8 July 2017

When terrorists or criminals need to move larger sums or require more complex financial arrangements for fund movement or laundering or both, they cannot simply rely on the physical transfer of funds. They need the wider range of services and tested transfer systems provided by the big retail banks and other large financial institutions of the formal and regulated banking system, which is itself the mainstay of many countries' financial systems. The speed, ease and efficiency of financial flows provided by these large-scale banking and financial institutions are undoubtedly attractive to both criminals and terrorists as is the aura of legitimacy and respectability that attaches to their involvement in the main gateway of commercial and retail financial transactions. FATF has noted: 'Combined with other mechanisms such as offshore corporate entities, formal financial institutions can provide terrorists with the cover they need to conduct transactions and launder proceeds of crime.'[17]

Examples of this include the case of Hizb-ul-Mujahideen, which received funds originating from Pakistan and made extensive use of the regional formal banking system to support its terrorist activities in India. The monies were employed to support active terrorists and also the families of terrorists killed in action. The Indian government claimed that funds were used to purchase mobile communication equipment, medical treatment, and arms and ammunition.[18]

The danger to both criminals and terrorists of using the formal banking sector is the legal and regulatory oversight to which users of the system will be subject. This is ultimately a significant deterrent to reliance on the formal sector by such parties. Important here too is that when weighing up the small amount of funds necessary for terrorist activities against the dangers of discovery as a result of regulatory systems and procedures (such as 'know your customer' checks and account monitoring), terrorists may seek to avoid the formal sector in its entirety.

While the 9/11 al-Qaeda terrorists did not shy away from the formal system and opened and used bank accounts, made wire transfers and employed travellers' cheques, since 2001 there has been a clear tendency for terrorists to shift to the physical transport of cash and reliance on the informal banking and money transmission sector. This has been in large part due to the increasing adoption of anti-money laundering (AML) and counter-terrorist financing (CFT) measures in the formal banking realm but it is also explained by the very limited amounts of funds needed in more recent attacks by smaller and more atomised groups. Indeed, Peter Neumann has argued that large amounts of terrorist financing never enter the formal banking system and that 'few of the financial transactions of terrorist groups appear on bank statements'.[19] Nevertheless, the US Treasury has pointed to the 'residual' risk of exposure to terrorist financing in the formal sector due to the size of global financial transactions that pass through the US financial system and particularly to the use of 'correspondent banking' networks.[20] This latter phenomenon is regarded by some analysts as an important factor in terrorist fund movement and has received significant attention from regulatory authorities with important consequences.

Correspondent banking can be defined as the provision of banking services by one bank (the 'correspondent') to another bank (the 'respondent') to enable the latter to provide its own customers with cross-border products and services which it is unable to provide itself, presumably because it lacks the international network or a particular product or service in the correspondent bank's jurisdiction. The correspondent bank acts as the agent of the respondent. The relationship can include trade-related business and also treasury or money market activities allowing for the transaction to be settled through the correspondent arrangement. The relationship can also involve establishing accounts, providing payment and clearing services, and exchanging methods of authenticating instructions.

As a simple example, say I wish to import fertilisers from Libya but my local bank in the United Kingdom (the respondent bank) does not have a branch in that country and so executes a payment transaction via a local correspondent bank to whom it forwards my money for onward payment in favour of the seller in Tripoli. Similarly, the Libyan bank may use the services of my local UK bank to carry out transactions in London because it lacks a presence in the United Kingdom. In this context the bank in the United Kingdom is acting as the correspondent and the Libyan bank as the respondent.

There are clear money-laundering and terrorist-financing risks built into this relationship. In the first instance it is often the case that the correspondent bank does not have a direct relationship with the parties to the transaction, the true identities of whom may only be known to the respondent. No significant customer due diligence may underpin the processing of wire transfers or clearing of cheques. The correspondent bank may choose to assume that the respondent bank has carried out these checks, though whether it has done so (and to what level) may be unclear. Essentially this type of banking is a non-face-to-face business for the correspondent bank.

Furthermore, and certainly prior to 9/11, little due diligence was carried out by correspondents and respondents on their partners in this banking relationship. Clearly, the financial and other attractions of functioning in a globalised market trumped fears of money laundering and terrorist penetration. In a study in 2011 by the United Kingdom's Financial Services Authority (FSA), one major bank (not named) reported that of its 2,500 correspondent-type relationships, 500 were closed down—400 on economic grounds (meaning that they were not really viable or readily used) but 100 were terminated on AML grounds.[21]

In the past, limited effort appears to have been spent on identifying high-risk customers and suspicious activities or prioritising high-risk customers and transactions for investigation in relation to correspondent activities. Poor practice included failure to consider money-laundering risks associated with correspondent relations and a lack of clear policies and procedures in dealing with banking partners in this kind of relationship, including partner customer on-boarding and procedures for approval of correspondent relationships. These failings were compounded by inadequate levels of assessment in respect of banking partner AML systems and controls, and consideration of the overseas AML regime within which a banking partner functioned. There was also at times a lack of ongoing monitoring and review mechanisms of the general correspondent relationship.

The FSA reported in 2011 on a wide range of compliance standards in respect of starting and managing correspondent relations. According to the FSA, one UK-based bank told it that customer due diligence files for all its respondent banks had been lost sometime before the FSA's on-site visit and that the head of correspondent banking at the bank could not account for how it managed its higher-risk respondent relationships.[22] In other instances, the FSA reported that one small bank had very little relevant information on correspondent banking relations on file and in another bank there had been no analysis of its respondents' AML procedures and policies because the London branch manager who had responsibility for opening correspondent accounts 'knew the banks personally'.[23] Elsewhere, a file review of a respondent bank located in an extremely high-risk jurisdiction revealed that one of its shareholders was a politically exposed person (PEP) but that this factor had not been identified by the bank. At another bank two PEPs had been identified but no action had been taken. Even more egregiously, at the latter bank screening of signatories appeared to have taken place just two weeks before the planned FSA on-site visit.[24] The FSA study also revealed that a major bank had identified a PEP who owned a majority stake in a respondent bank and had been implicated in a large bribery scandal but that the risk manager had concluded: 'I suggest to keep the relationship.'[25] One major bank maintained a correspondent account owned by a respondent who had been the subject of approximately 300 suspicious activity reports in two years. Nevertheless, the account was retained because the respondent's parent was a government-owned entity and the bank felt that if it terminated the relationship, it would affect its ability to operate within the respondent's jurisdiction.[26]

Over recent years the United States in particular has become concerned about the dangers of so-called 'nesting' in correspondent accounts. This is a system whereby a foreign financial institution sets up a correspondent account in a US bank and permits a third party to have access via one of its accounts to an account in the United States. Potentially, of course, that third party could be a criminal, a terrorist or a sanctions breaker. These individuals or organisations will thereby gain anonymous access to the US banking system and be able to conduct a range of financial transactions including currency exchange, loans, and the clearing of cheques and money. Such nesting activities in turn may involve money laundering and terrorist fund movement.[27]

The vulnerability of correspondent banking to both criminal and terrorist funding penetration and nesting led to a growing focus on this aspect of the banking system. In the United States, very high fines were imposed upon banks which did not monitor their correspondent relationships too closely. In 2012, for example, a US bank paid a $1.92 billion fine to the authorities because it was used to launder drug money emanating from Mexico, while in 2015 a German-based bank and its US subsidiary were fined $1.45 billion for breaking laws forbidding transactions with Sudan, Cuba and Myanmar. In both instances US authorities referred to violations of the 1970 Bank Secrecy Act including failure to have an effective AML programme, failure to conduct adequate due diligence, failure to report evidence of money laundering or to obtain 'know your customer' information with respect to foreign correspondent bank accounts, and failure to detect and adequately report evidence of money laundering and other illicit activity.[28]

FATF also expressed concern about the dangers posed by correspondent banking. It noted in 2016: 'Jurisdictions should also protect against correspondent relationships being used to bypass or evade counter-measures and risk mitigation practices', especially in respect of North Korea.[29] Its guidelines certainly paid significant attention to the dangers of correspondent relationships.[30] Relevant FATF recommendations, especially nos. 1, 6, 7, 10, 11, 13, 14, 16, 20 and 26, touch upon correspondent banking issues as do many of its other reports. (We shall review several key FATF recommendations in the next chapter.) In addition, correspondent banking received attention from governments and a range of other financial institutions involved in regulating global financial practices.[31]

Together with massive fines imposed by the US authorities on those financial institutions that fell short of proper relevant safeguards, and new regulations in respect of capital and liquidity that emerged after the 2008 financial crisis, the global focus on the vulnerabilities of correspondent banking pushed banks to start severely restricting their correspondent practices. The formal sector's move to curtail correspondent relations soon turned into a rout as banks saw it was in their immediate interest to totally avoid rather than manage the attendant risks.

The case of Latvia is illustrative of the speed of the de-risking process. Following the break-up of the Soviet Union, Latvia sought to use its banking system to become the 'Switzerland of the East' by attracting deposits from citizens of Russia and the former Soviet republics and establishing itself as an offshore banking centre for Russian corporates.[32] Little due diligence and rudimentary AML procedures soon attracted the attention of US AML and CFT organisations, and the US Treasury accused the Latvian banking system of 'institutionalised money laundering and helping fund North Korea's nuclear and missile programme'.[33] In 2013 JPMorgan Chase withdrew from the country, soon followed by all major international banks—the last, Deutsche Bank, exiting in 2017.[34]

Latvia was soon generally devoid of foreign banks—a major blow to that country's aspirations to integrate into the European Union and global economies. But this is not a problem unique to Latvia. A report by the Financial Stability Board in 2017[35] demonstrated that de-risking was a global phenomenon, with the most affected regions being the Caribbean, Eastern Europe, Central Asia and East Asia. Eastern Europe witnessed the greatest decline in correspondent relations (20 per cent) followed by the Caribbean (10 per cent).[36] Products and services most impacted include international wire transfers, cheque clearing, clearing and settlement, and cash management services. Customers most badly hit include local small and medium domestic importers and exporters, money transfer operators, and other money service businesses, including remittance providers.[37]

Ironically, as banks de-risk partly in response to fears of money laundering and terrorist financing, the termination of correspondent banking relations may actually serve to abet both practices. This is because entire jurisdictions are now being excluded from the global financial system; and in a country such as Somalia, where 40 per cent of the local populace are critically dependent upon remittances from abroad, the result can be catastrophic.

Following the termination of all US banking facilities for providing remittances to Somalia, Somali Americans began a Twitter campaign employing the hashtag #IFundFoodNotTerror.[38] The consequence of preventing countries such as Somalia and Afghanistan from accessing the global financial system is that citizens of those states wishing to transfer funds will increasingly be forced to turn to the informal financial sector, where—as we shall see in the following section—regulatory and legal oversight of financial transactions is severely lacking compared to that found in the formal and regulated financial realm. Not only will intelligence about such transfers and participants inevitably decrease, but both criminals and terrorists may be better able to disguise and launder their funds among the now larger pool of informal banking participants.

In addition, where financial instruments are used more offensively in support of foreign and military policy, de-risking may have a negative impact too. Zarate thus argues that 'forced isolation of more and more actors—and the tendency of the private sector to decline doing business in at-risk sectors and jurisdictions and with suspect actors—raises the possibility of reaching a tipping point where the effectiveness of these tools begins to diminish'.[39] This will occur because the leverage provided by the formal financial sector will diminish as its presence retreats in areas where it could use its influence to support broader political goals.

Recognition of the negative impact of de-risking for AML and CFT has led the FSB, the International Monetary Fund, the Basel Committee on Banking Supervision, FATF, the World Bank and other organisations to put forward various suggestions about how to better address the consequences associated with the decline of correspondent banking and also perhaps nudge financial institutions towards better managing their correspondent relations rather than terminating them entirely. These include clarifying regulatory expectations (including new guidelines by FATF in the application of AML and CFT to correspondent banking especially in respect of expectations around customer due diligence in high-risk scenarios, remittances, financial inclusion and non-profit organisations);[40] capacity building in countries that are home to the affected respondent banks, with the aim of reducing the problems that gave rise to de-risking in the first place; and enhancing tools for due diligence by correspondent banks, including information sharing and 'know your customer' systems. Whether such efforts will actually encourage large numbers of banks to return to the correspondent market is questionable given the continued aggressive approach of the US authorities towards AML, CFT and sanctions infringements.

The Informal Banking System: Money Service Businesses and Hawalas

Imagine a Bangladeshi construction worker in Qatar engaged in work on the World Cup stadiums. Every so often he needs to transfer funds back to his family in Dhaka. Or consider a parent in the United Kingdom wishing to send money to their gap-year teenager short of funds in Phuket, Thailand, or an American parent forwarding cash to their spring-break teenager in Cancún, Mexico. In these cases the process will involve supplying the money to a transfer provider intermediary who will then forward the funds to the intended recipient overseas. The funds can be transferred to the intermediary by way of a mobile app online or over the phone. The Bangladeshi worker or UK-based parent can use cash, debit or credit cards, or they can move money from their bank accounts to the intermediary, who will then ensure that the funds are ultimately deposited in the recipients' bank account or mobile wallet or made available to them in cash. Options for transferring the money overseas are diverse and include various types of intermediaries, including bank to bank account transfers, money transfer companies with cash pick-up options, online systems such as PayPal, and international cheques. Transfer fees, exchange rates and minimum transfer amounts may vary significantly. There are other considerations that both sender and recipient need to take account of, including pick-up methods, turnaround time and transfer options, such as reductions for recurring payments and the availability of forward contracts whereby the sender may lock in an exchange rate for future transfer.

What is common to both the Bangladeshi worker and the American or British parent in these scenarios is that the amounts they may wish to transfer are generally small yet the potential costs of even simple transfers are often not insignificant. The costs of services in terms of fees and exchange rates provided by the formal banking systems (with their large overheads) may make such small-time transfers far from economical. Even if their rates are competitive, the services provided by these institutions may still be less than totally flexible—assuming they can actually provide such services in areas in which the intended recipient is located.

The point here is that because of a combination of costing issues, lack of service provision, and convenience, some people, including migrant workers, residents of war zones and ungoverned spaces, or simply those wishing to reduce transfer costs, may be forced to gravitate to money transfer systems and money service businesses. These services have in many (though not all) cases been less subject in the past to regulatory oversight than the formal banking sector, and in many (though again not all) cases have been characterised by weaker and less transparent record-keeping and the possibility of greater anonymity. It is these characteristics that make the system vulnerable to terrorist and criminal penetration as they may be employed to evade currency controls, tax obligations, sanctions and general regulatory oversight.

The informal sector—by which we mean that part of the financial sector separate from the main institutional banking world—consists of a range of diverse institutions with varying primary roles and both historically and contemporaneously subject to differing levels of regulation. Services provided by this sector include money transmission services, cheque cashing and foreign exchange facilities. These providers range from multinational corporations such as Western Union, which are subject to regulatory oversight, to services provided by small and independent local shops, which may not always be so regulated.

One form of money service business that has presented problems for AML and CFT systems are the Hawalas, which have often played an important role among migrant communities working in global financial centres or far from their homes. Originally developed in India, Hawala is Arabic for transfer or remittance. Hawalas and other related service providers often have links to particular geographic areas or ethnic communities. They generally operate in areas with a high percentage of expatriate workers and in the vast majority of instances provide totally legitimate and necessary services for those who cannot use the formal banking system for the cost and convenience reasons outlined above.[41]

For Hawala operators the money transfer and currency exchange business is in some cases often part of other business operations, which could include import and export companies or simple convenience stores. Often working on the back of family connections and a system of trust settled on a net basis, money has traditionally been transferred without customer due diligence, or paper trail documenting key features of the transaction,[42] or any reporting requirements to local authorities. This is made possible by the fact that these organisations deal mainly in cash and in personal remittances of generally low value. The Hawala and other similar money service businesses involve in their transactional practices the communication of very limited information, mostly basic information on both customer and beneficiaries, contact numbers and sometimes a transaction reference number.

The scale of these services is therefore difficult to quantify but Pakistani officials have estimated that in excess of $7 billion flows into their country through Hawala networks each year.[43] In a FATF study of Hawalas and related service providers, only a limited number of countries responded with estimates of the scope of their domestic unregulated sector and, when they did do so, the estimates were not particularly precise, ranging from 25 to several hundred such organisations.[44]

In view of the lack of effective regulatory oversight, patchy record-keeping and reporting procedures, the relative anonymity potentially available to these kinds of transactions, and their connection in some instances to zones of conflict, FATF has focused considerable attention on this type of business. In its studies FATF has identified three types of Hawalas and other similar service providers: legitimate Hawalas, which provide legal money transmission and currency conversion services (it should be noted that they represent the vast majority of Hawala services); hybrid traditional Hawalas, which may unwittingly be exploited for purposes of criminal money laundering and terrorist financing; and criminal Hawalas, which are directly complicit in money laundering and terrorist financing.[45]

Criminal Hawalas and similar service providers facilitate the transfer and laundering of proceeds from drug trafficking, fraud and smuggling. In relation to terrorist financing, they are vulnerable because of the lack of any supervisory function, the use of net settlement, and the mingling of illegal and legal proceeds. As FATF notes: 'Inadequate efforts of outreach to the unregulated sector to pull them into the regulated sphere in some countries plus limited or no enforcement actions against unregistered entities also minimizes the incentives for unregulated entities to subject themselves to regulation and supervision, making them more vulnerable to terrorist abuse.'[46]

There are indeed several examples of Hawalas being used to support terrorist operations. In 2011, Mohammad Younis was charged in US courts with operating an unlicensed money transfer business between the United States and Pakistan. One of the transfers was used to fund the 2010 attempted car-bombing in New York City's Times Square. In the so-called Carnival French Ice Cream case, Abad and Aref Elfgeeh ran a money transfer business in Brooklyn which was involved in the wiring of small amounts of money to accounts in 25 countries, including to an al-Qaeda operative in Yemen. In 2008 another Pakistani national residing in Washington DC was found guilty of conspiring to use a remittance system to support international drug traffick-ing and smuggling of counterfeit cigarettes and weapons as well as providing assistance and financing to members of al-Qaeda.[47]

The UK authorities in turn have always faced a dilemma in respect of Hawalas and other money transmission and currency exchange services. World Bank research in 2016 noted that remittances originating in the United Kingdom have played an important role in supporting overseas economic development: funds proceeding to Nigeria were estimated at $3.8 billion in 2014 while India received $3.7 billion in remittances that same year.[48] Efforts to stymie or control this network would certainly have a negative impact on funds transferred to developing economies: these totalled about $11.5 billion in 2014. Nevertheless, the UK authorities have identified money service busi-ness as high risk—up from medium risk in the 2015 assessment—with some such services being controlled by criminal organised groups. In addition, the sector was identified as a high risk for terrorist financing because of its connec-tion to high-risk jurisdictions and the lack of appropriate supervision.

The UK's 2017 *National Risk Assessment of Money Laundering and Terrorist Financing* stated that there were approximately 2,000 money service busi-nesses registered in the United Kingdom (a reduction of 1,000 since 2015), which included 1,000 firms providing money transmission services, 500 pro-viding cheque-cashing services and 1,300 providing currency exchange ser-vices. A significant finding was that while only 2,000 principals were registered, money business services were being conducted in over 45,000 premises through principals and, crucially, their agents.[49]

This principal–agent model is a key characteristic of these money transmis-sion services. Significantly, as formal banks have de-risked and pulled out of relationships with the informal sector, there has been a decrease in the number of registered principals who are responsible for their agents' due diligence procedures. As a result, supervision in respect of AML and CFT requirements has suffered a decline.[50] UK authorities claim to have noted that as a result of de-risking and lack of access to the formal banking network, the money trans-mission services have started altering their business model, and with greater emphasis now on anonymous transactions, there has consequently been greater use of the services by those seeking to launder criminal funds. Evidence of this includes the growing use of courier services or parcel and freight companies to send cash out of the United Kingdom. In one case a money transmission agent sent £14,000 in suspicious cash to an overseas destination after previously sending 70 parcels to the same country.[51] Other examples involve the use of illegal Chinese immigrants brought into the United Kingdom by Chinese criminal groups, who were then employed to work in money service businesses involved in criminal money laundering in order to pay off their debts to the traffickers. Albanian drug-trafficking gangs also set up money service businesses to convert sterling into higher-denomination euro notes for purposes of transfer to the continent. In Northern Ireland, currency exchange businesses operating close to the Irish border have been important mechanisms for laundering funds.[52]

In the United Kingdom money service businesses are also vulnerable to penetration by terrorist financiers. A key characteristic here is employees who are complicit in transferring funds to terrorist organisations. In 2014 two individuals based in the United Kingdom used a money service business to transmit funds to a third party in Turkey for forwarding to a family member fighting for ISIS in Syria, while in Northern Ireland money service businesses have been employed to move the proceeds of VAT evasion or fraud to support terrorist groups.

As noted above, FATF regards the lack of adequate supervision as a crucial variable in the vulnerability of this sector to criminal and terrorist penetration. The UK authorities appear to take a similar position: HM Revenue and Customs has increased its supervisory activities in respect of money service businesses.[53] Poor practices within the sector have been identified as including inadequate investigations by principals of their agents, inadequate due diligence, and poor record-keeping and poor monitoring of transactions. The UK government is also concerned that de-risking in the formal sector will mean increasing movement to this informal and relatively far less regulated sector, and it is therefore keen to increase remittance providers' access to formal banking services.

The problem in the United States is not too dissimilar to that in the United Kingdom though of course there are differences of scale. As noted in the US *National Terrorist Financing Risk Assessment 2015* (*NTFRA 2015*):

> The [money service business] industry in the United States is extremely diverse, ranging from Fortune 500 companies with numerous outlets and agents world-

wide to small, independent 'mom and pop' convenience stores in communities with population concentrations that do not necessarily have access to traditional banking services or in areas where English is rarely spoken.[54]

The *NTFRA 2015* differentiates between licensed money service providers and unlicensed ones and notes that problems are, unsurprisingly, potentially more acute in the latter group. The Younis case mentioned above is an example of the dangers of unlicensed money services and the US authorities have sought co-operation with financial institutions to better detect and report these unlicensed activities.[55]

While Hamas and Hezbollah have in the past preferred the formal banking sector in the United States, al-Qaeda affiliates and al-Shabaab have shown a proclivity for employing money service businesses. Examples of this include individuals raising funds for al-Shabaab from within Somali communities in Missouri and then transferring the funds via licensed money service businesses. There have also been cases of individuals using such licensed services to send money to the Taliban.

In the United States money service principals (though not agents) are required to register with the relevant authorities. Furthermore, the 1970 Bank Secrecy Act requires money service businesses to file currency transaction reports and suspicious activity reports and also to maintain various records. These businesses are also required to adopt reporting, record-keeping and customer identification measures. The majority of states require money service businesses to be licensed within the state.[56]

Prepaid Cards and Internet Payment Services

So far we have looked at particular sectors that may play a role in the transmission of criminal and terrorist funds, but what about products? Here we will review several products offered by banks and corporate organisations as well as various internet services which may be vulnerable to criminal and terrorist exploitation.

Reloadable stored value cards, also known as prepaid debt cards, come in two kinds: closed-loop and open-loop systems, with the latter being potentially most at risk of malevolent exploitation. Closed-loop cards usually restrict the user to buying goods or services from the merchant or service provider issuing the card. They are also sometimes used for public transportation or within a university campus (or other such closed institution) to pay for food or other services. Value can be added to the card but it can only be employed in a restricted context, thus limiting its attractions to criminals or terrorists, who may require a means for moving money across borders and also employing the cards in wider environments.

On the other hand, open-loop systems may carry the brand of a payment network (such as American Express or Visa) and can be accepted by a very wide range of merchants and at ATMs, which connect to a global payment network.

Money-laundering vulnerabilities of open-loop products manifest in several ways. Criminals may use such cards to cash out the proceeds of fraud while terrorists may use them to transfer funds, including across borders. Thus criminals and terrorists can load the cards in one jurisdiction and inconspicuously carry them across borders, and then in the target jurisdiction the card value is converted back into cash, the amounts only being restricted to ATM withdrawal limits.

Provided the amount that is loaded onto the card is below AML and CFT limits, there is usually no significant customer due diligence or reporting that takes place. Multiple cards can be linked to the same account, and it is difficult to determine the identity of the card user as anyone can employ it provided they have the pin number. The 2015 Bataclan attack and associated atrocities in Paris, for example, were partly funded using prepaid cards. The prepaid cards, some of which were purchased in Belgium, were used to pay for cars and apartments for the terrorists in the 48 hours preceding the 13 November attacks.[57] Obviously, cash limits on uploads and withdrawals may make these products strategically more useful to terrorists than high-end money launderers.

Payment services based on the internet offer further options to terrorists seeking to move funds. ISIS officials have reportedly used eBay and PayPal to funnel money to an alleged operative located in the United States.[58] In respect of these services, FATF has noted:

> Approximately half of all terrorism financing suspicious transaction reports concern customers aged between 21 and 35. The use of an online payment system to assist in financing terrorism is more a reflection of the prevalence of this payment system in the wider financial system rather than any indication that online payment systems are more vulnerable to terrorism financing.[59]

Nevertheless, in view of the potential anonymity and efficiency of these payment mechanisms, they cannot be ruled out as vulnerable to criminal and terrorist exploitation.

Emerging Threats: Cryptocurrencies and Blockchain

In early October 2017, as part of a training exercise, I advised my Finance and Security students in the Department of War Studies to buy Bitcoin. Those that did and then held on to the cryptocurrency until the end of the semester in December would have doubled their investment. It was the best investment advice I have ever given. The cryptocurrency (especially Bitcoin) frenzy of that year raised public awareness of the attractions of various digital currencies (there are hundreds of types of decentralised digital currencies) though few fully grasped its workings or initially cared about the potential dangers that this new technology posed for money laundering and terrorist financing.[60]

Major advantages of decentralised cryptocurrencies—a subset of virtual currencies—include the efficiency and speed of transactions, its immutability (once the transaction is initiated it cannot be stopped), the avoidance of expensive fees, and the attendant anonymity. Central to these attractions is the fact that cryptocurrencies such as Bitcoin incorporate cryptographic principles linked to a distributed public ledger system known as blockchain technology. This technology permits peer-to-peer transactions external to the mainstream financial system where AML and CFT energies are concentrated. The creation of currencies, the validation of the transaction and the actual payments are carried out between the parties directly without third-party monetary authority or regulatory oversight.

If I wish, for example, to pay a counterparty in Egypt $20,000, I would normally go to my local bank, which, before transferring the money, would subject me to 'know your customer' verification procedures and various other computer and manual checks to reduce the risk that my transaction is illicit. Should the transaction raise suspicions, not only would the transaction be frozen but a suspicious activity report would be submitted to the relevant authorities warning them that something amiss was possibly taking place. At least in theory that is what is supposed to happen.

Alternatively, if I wish to use Bitcoin or other products such as Litecoin or Ethereum to pay my Egyptian counterparty, I announce the transaction over the internet and use various passwords with both public and private elements, thereby initiating a validation process by millions of computers located globally. My Egyptian transaction is recorded and time-stamped using a hash algorithm and together with other newly initiated transactions are placed into a block. And this new block is then connected to previous transactions in a chronological blockchain. My transaction is thus verified by millions of independent global users rather than by a single intermediary.

I pay no intermediary fee but no one checks my transaction for legality. Most significantly, both sending and receiving Bitcoins can be undertaken without revealing any identifying information. Digital wallets employed to store Bitcoin do not require me to provide information as to my identity, and should I use multiple electronic addresses for each transaction, the identification process becomes even more challenging. The random transmission of the transaction over the peer-to-peer system makes it very difficult to ascertain where the deal originated. Techniques using mixing services known as tumblers, which aggregate transactions from multiple users, only add to the opacity of the process.

Clearly, the decentralised nature of digital currencies establishes them as particularly resistant to law enforcement and government control thereby making these currencies extremely attractive money-laundering instruments. Potentially, by relying on the anonymity provided by the system, criminals and terrorists can hide their true identity as they move, layer and then integrate digital currencies. The problem here is that key AML and CFT techniques intrinsic to the preferred risk-based approach (which we will look at in greater detail in Chapters 4 and 5), such as 'know your customer' systems and account monitoring systems (and ultimately suspicious transaction reports), cannot be employed in the cryptocurrency space. Here the placement, layering and integration phases of money laundering are speedily accomplished with anonymity and no oversight. Concomitantly, the solicitation and transfer of funds by terrorists benefit from a similar environment.[61]

There are several examples of criminals using Bitcoin, including most notably the Silk Road, a digital market for illegal drugs which accepted Bitcoin payments and whose owner accumulated tens of millions of dollars in Bitcoin, which he proceeded to launder.[62] There are many cases of criminals exploiting the anonymity of cryptocurrencies in the Netherlands, China and California. According to one estimate, between $80 billion and $200 billion of cybercrime-generated cash is laundered every year.[63] In 2017 Europol noted that Bitcoin remains a key facilitator for cybercrime, though other cryptocurrencies such as Monero, Ethereum and Zcash are becoming increasingly popular.[64]

In academic research sponsored by Bromium Inc., it was found that virtual currencies are the primary tool employed by cybercriminals for money laundering and that there was already a process to move away from Bitcoin to Monero, which provides even greater anonymity. Report author Mike McGuire commented:

It's no surprise to see cybercriminals using virtual currency for money launder-ing. The attraction is obvious. It's digital, so is an easily convertible way of acquir-ing and transferring cybercrime revenue. Anonymity is also key, with platforms like Monero designed to be truly anonymous, and tumbler services like CoinJoin that can obscure transaction origin.[65]

A somewhat alternative view is provided in the 2016 *National Strategic Assessment* produced by the UK National Crime Agency, which claimed that 'virtual currencies have yet to be adopted to any large degree by money laun-derers'.[66] Of course things may change rapidly, but the differences in outlook here are nevertheless quite significant.

While criminals may have an interest in this medium, what about terror-ists? FATF certainly believes terrorists can exploit the anonymity of the sys-tem and in 2015 identified virtual currencies as a risk in relation to terrorist financing.[67] However, so far the number of actual uses of cryptocurrencies ascribed to terrorists has been minimal. There have been a few cases. In 2015 Ali Shukri Amin, who had published an article entitled 'Bitcoin and the Charity of Jihad', was sentenced to eleven years in prison for providing finan-cial support to ISIS. He employed a Twitter account to publish instructions as to how donors could use Bitcoins to hide their identities and transfer funds to ISIS and how they could support foreign fighters seeking to reach the war zones in Syria and Iraq. In 2016 a former CIA agent identified the Ibn Taymiyyah Media Center, a Gaza-based online jihadist news agency, as having received Bitcoin donations.[68] In 2017 the Indonesian authorities announced that an ISIS agent who was then active in Indonesia had engaged in Bitcoin transactions with other jihadists in support of terrorism.

Why are the cases so limited? First, the setting up of a wallet account and the use of and employment of a cryptocurrency are quite complex, and this complexity can act as a deterrent to use. Not all criminal or terrorist groups have the technical capability to manage cryptocurrency effectively and effi-ciently. Secondly, the system is not totally anonymous: it may be possible to link a user to his alphanumeric key and also to cluster multiple cryptocurrency pseudonyms and link them to the user. It needs to be borne in mind that blockchain technology creates a record of the transactions underpinning the movement of funds: indeed, the Silk Road was unravelled to a degree because of its exploitation of Bitcoin. Anonymity can be further tested at the point of integration when it comes to exchanging the virtual currency for fiat currency (legal tender) as identification will be required by gatekeepers. Criminals and money launderers will not view such challenges with equanimity.[69]

The value of Bitcoin may also alter quickly and dramatically, and an individual's wallet may be hacked and stolen, not a situation that either criminals or terrorists can readily tolerate. Nor is the cryptocurrency market very liquid, and there is always the possibility that a cryptocurrency holder may not be able to locate parties willing to exchange cryptocurrency for fiat currency. To all this must be added the increasing interest regulators and governments are showing in cryptocurrencies, and the uncertainty that cryptocurrency users may have about the future efficacy of law enforcement and technical efforts.

But perhaps most significant is the fact that given the small amount of funds terrorists require, there is little incentive for them to transfer from other, more tested methods described above. Cash and its alternatives like prepaid cards are not perfect but remain highly efficient means by which many terrorists can transfer the limited funds they require for their actions. The danger of terrorist employment of cryptocurrencies may still arise, but at the moment it represents more of a theoretical threat than a real one.

As David Carlisle writes: 'It is not yet clear whether cryptocurrencies will become a major terrorist funding tool, at least in the near term, and the longer-term picture remains uncertain. Indeed, terrorists already have a number of reliable financing streams, which show little sign of drying up.'[70] Carlisle goes on to argue: 'Treating cryptocurrencies as an exceptional threat creates the misleading impression that more conventional financial products are not already equally, or more, vulnerable to terrorist exploitation.'[71] Currently, while the United Kingdom authorities continue to recognize the inherent vulnerabilities of virtual currencies, they regard the associated risk of money laundering and terrorist financing as low.[72]

Thus for both criminals and terrorists the movement of funds presents a distinct set of challenges distinguishable from the hurdles that need to be overcome in the generation of funds. Criminals move funds to pay off liabilities but ultimately to disguise the funds' origins and to make them readily available for enjoyment in safe environments. For terrorists the goal is to move the funds from where they are generated—often in relatively wealthy and stable financial areas—to regions where they are to be employed, whether war zones or closer to the source of generation should the planned terrorist attack be local.

One of the key differences between the two groups is the amount of funds required to be transferred. Where large amounts need to be laundered, the processes of placement, layering and integration may require the complex formal banking system of large banks and sophisticated financial institutions.

The myriad services and products available from this sector make it attractive or at least ensure that its attractions will be balanced against the dangers of regulatory oversight that come with any reliance on this system. Terrorists may similarly choose to use this network, but the relatively small amounts of funds required—at least in many of the recent attacks in Europe and elsewhere—means that the dangers of the regulatory system may be avoided in their entirety by physically transporting the funds to the intended target.

One cannot underestimate how the increased regulatory pressures on the formal sector have led to the process of de-risking, which has had the ironic effect of enhancing the role of the informal financial sector, parts of which fall well outside any effective regulatory ambit. Both licensed and unlicensed money service providers have proven vulnerable to penetration by criminals and terrorists, and some means need to be found to extend better regulatory oversight to this subsector. Products such as prepaid cards and internet payment services also need fuller scrutiny though, at the time of writing, cryptocurrencies, because of their complexity and uncertain transparency, have yet to emerge as a serious terrorist financing threat. But while the movement of funds may be readily distinguishable from the requirements of their generation, both fund generation and fund movement continue to be targeted by global and national AML and CFT structures and policies.

4

GLOBAL RESPONSES AND THE RISK-BASED APPROACH

The international financial system is only as strong as its weakest link.

FATF, 2014[1]

The Global Level

In 2017, after three years of war with Russian-backed separatists in the east of the country, the Ukrainian government requested the International Court of Justice in the Hague, the United Nations' highest court, to impose fines on Russia for intervening militarily in Ukraine, for violating the human rights of Ukrainian citizens but also for the specific offence of financing acts of terrorism. In this regard, the Ukrainians accused Russia of violating the 1999 International Convention for the Suppression of the Financing of Terrorism by supplying money, weapons, training and other support to separatists in the self-proclaimed Donetsk and Luhansk People's Republics in Eastern Ukraine.[2] Reliance on this UN convention aimed at tackling terrorist financing by one nation state against another in the midst of a war was novel indeed. The Ukrainians' argument would, however, at some point inevitably run up against the problem that has long bedevilled UN approaches to the problem of terrorism—definitional debates as to what constitutes a terrorist. Nevertheless, the submission indicated that measures aimed at countering terrorist financing could be weaponised in many different ways, and not necessarily in the manner originally conceived by their authors.

While the implementation of anti-money laundering (AML) and counter-terrorist financing (CFT) measures takes place at the national level, the driving force behind AML and CFT strategic policy and standards is to be found among global institutions. Given the global nature of the problems and the critical requirement of co-ordinated international action, this top-down approach is understandable. The Financial Action Task Force (FATF) is the key international inter-governmental organisation which establishes global principles for AML, CFT and proliferation finance and presents them to local jurisdictional authorities which, in turn, attempt to impose consistent standards on domestic financial and non-financial institutions and other relevant parties. At the core of FATF's methodology is its support for the adoption by states and institutions of a so-called risk-based approach (RBA). This key intelligence tool is critical to both national and institutional responses to a range of threats posed to financial security and stability.

But FATF is not the only global body involved in the protection of the integrity of the global financial system. The United Nations and its various agencies also establish conventions and pass resolutions to combat challenges to international financial security, and UN member states are required to act on them. And, of course, intermediaries such as the European Union and other regional bodies also interpose between the global and national levels for purposes of co-ordinated local implementation.

The United Nations and FATF work to reinforce each other, though it should be noted that the UN approach to AML and CFT is part of a broader counter-terrorist strategy while that of FATF is specifically targeted against criminal money laundering and terrorist financing as ends in themselves. It should also be noted that the UN approach predated that of FATF, and it is therefore to the United Nations that we first turn.

The Evolution of the United Nations Approach to AML and CFT

The United Nations had been concerned about the distinct problems of terrorist financing and also criminal money laundering and organised crime well before the crisis in Ukraine and even before the terrorist attacks of 9/11. Approaches to criminal money laundering and terrorist financing were, however, initially not integrated and effectively proceeded in parallel. The financial challenges presented by terrorism were viewed as part of the broader political and security threats posed by a growing terrorist phenomenon, while anti-money laundering strategies developed as part of a response to the challenges of the global drugs trade.

The origins of the United Nations approach to AML and CFT are thus to be found in counter-drug measures adopted in the early 1960s. In 1961 the Single Convention on Narcotic Drugs (as amended by the 1972 Protocol) was adopted; it was aimed at prohibiting the production and supply of specifically defined drugs unless produced under licence. This was followed by the 1971 Convention on Psychotropic Substances, which aimed at international agreement for controlling various psychoactive drugs. This latter convention was in turn complemented in 1988 by the United Nations Convention against Illicit Traffic in Narcotic Drugs and Psychotropic Substances (the Vienna Convention), which sought to combat organised drug crime and required the tracing and seizing of drug-related assets. These three multilateral treaties dealing with the control of drugs are mutually supportive and serve to reinforce each other in terms of objectives and guidelines.[3]

Then in 2000, the UN General Assembly began to broaden its focus on drugs to more directly address the issue of global crime by adopting the Convention against Transnational Organized Crime and its protocols (the Palermo Convention), which called for global action on the trafficking of persons, the smuggling of migrants, and the manufacturing and trafficking of firearms; it also included within its scope an attempt to address critical financial and criminal money-laundering components underpinning these threats. The convention involved the requirement for member states to adopt a legal framework to combat international criminal enterprises, to engage in global co-operation, and to undertake training and technical assistance.[4]

At about the same time, the United Nations also began developing several international legal instruments aimed at preventing acts of terrorism, instruments which in their financial aspects ultimately began to merge with approaches to international crime.[5] Soon after the 1998 al-Qaeda attacks on American targets in Kenya and Tanzania, which left over 200 dead and many more wounded, the UN focus on the drug trade and criminal money laundering was strengthened by the adoption during the following year of the International Convention for the Suppression of the Financing of Terrorism (the Terrorist Financing Convention). The Terrorist Financing Convention promotes co-operation aimed at preventing, investigating and punishing terrorist financing. Signatories to the convention commit themselves to the freezing and seizure of funds intended to be used for terrorist activities, which are then to be shared with all state parties. Moreover, state parties commit themselves not to use bank secrecy as a justification for refusing to co-operate in the suppression of terrorist financing,[6] a problem that still continues to bedevil

AML and CFT. In that same year, 1999, the United Nations Security Council, similarly motivated by the East African terrorist attacks, passed Resolution 1267, which, among other things, designated Osama bin Laden and his associates as terrorists and called for financial sanctions against both al-Qaeda and the Taliban.[7]

Following the al-Qaeda terrorist attacks of 9/11, the United Nations then intensified its actions against terrorism as well as both terrorist financing and criminal money laundering, and increasingly combined approaches against the latter within its structures and policies. Furthermore, UN policies now more explicitly aimed at supporting those of FATF, with the United Nations essentially allowing the FATF the lead in generating relevant international guidelines.

Consequently, in the immediate aftermath of the 9/11 terror attacks, acting under Chapter VII of the United Nations Charter, the Security Council adopted Resolution 1373 (2001), which included a strong financial component and which, among other proposals, called upon states to criminalise the financing of terrorism, freeze funds related to persons involved in terrorism, deny all forms of financial support for terrorist groups, suppress terrorist safe havens, share information with foreign governments, and criminalise active and passive assistance for terrorism.

Related Security Council resolutions followed rapidly, including 1535 (2004), 1540 (2004), 1566 (2004), 1624 (2005) and 2178 (2014), all aimed at strengthening the global counter-terrorist regime, including countering terrorist financing. By 2005 efforts were explicitly being made to align UN approaches with that of FATF. Thus Resolution 1617 (2005), which again targeted al-Qaeda with financial and other measures, urged member states 'to implement the comprehensive, international standards embodied in the Financial Action Task Force's (FATF) Forty Recommendations on Money Laundering and the FATF Nine Special Recommendations on Terrorist Financing'.[8]

At the same time, Resolution 60/288 of the General Assembly passed in September 2006 adopted a Global Counter-Terrorism Strategy whose four pillars involve addressing conditions conducive to the spread of terrorism, calling for the combating of terrorism, increasing national counter-terrorism capacity, and ensuring respect for human rights and the rule of law. The strategy also includes a call on states to refrain from financing terrorism.[9] The strategy aims (in Section II, paragraph 10) 'To encourage States to implement the comprehensive international standards embodied in the Financial Action Task Force's Forty Recommendations on Money-Laundering and Nine Special

Recommendations on Terrorist Financing of the Financial Action Task Force, recognizing that States may require assistance in implementing them.'[10] Also (in Section III, paragraph 8) the strategy seeks 'To encourage the International Monetary Fund, the World Bank, the United Nations Office on Drugs and Crime and the International Criminal Police Organization to enhance cooperation with States to help them to comply fully with international norms and obligations to combat money-laundering and the financing of terrorism.'[11]

While resolutions in both the Security Council and General Assembly attempted to set goals, norms and guidelines for member states to follow, already in 1997 the United Nations began establishing a group of entities and agencies responsible for implementing policy. Thus the United Nations Office on Drugs and Crime (UNODC) was tasked with leading and managing operational activities against drug trafficking, a role that was by 1998 widened to cover all organised crime and money laundering and soon thereafter also incorporated the campaign against terrorist financing as well.

A focus on entities and agencies specifically dedicated to counter-terrorism goals gained momentum after the 9/11 attacks. Security Council Resolution 1373 (2001) established the United Nations Security Council Counter-Terrorism Committee (CTC), which aims to support members in the prevention of terrorism, and through its executive directorate conducts national counter-terrorism assessments and provides counter-terrorism assistance to member states.[12] At the time of the adoption of the Global Counter-Terrorism Strategy, the General Assembly also established the Counter-Terrorism Implementation Task Force (CTITF) to co-ordinate UN efforts in respect of counter-terrorism, and in 2011 the CTITF was complemented by the United Nations Counter-Terrorism Centre (UNCCT), which provides capacity-building assistance to member states in implementing the global strategy. The importance given by the United Nations to these two organisations was reflected in the fact that in 2017 both were placed under the auspices of a new United Nations Office of Counter-Terrorism headed by an Under-Secretary General, which seeks to enhance the efficiency of UN counter-terrorism processes.

The CTITF consists of 38 international entities together with Interpol. It oversees twelve thematic working groups, which build capacity among member states in support of the Global Counter-Terrorism Strategy and includes a working group on Countering Terrorist Financing. In 2009 the working group prepared a report covering five key issues: the criminalisation of terrorist financing; the enhancement of domestic and international co-operation;

money transfer systems; non-profit organisations; and the freezing of assets. On the basis of these identified issues, the International Monetary Fund helped prepare an action plan for implementation by members of the working group. In 2013, the working group started a global capacity-building project for training officials of member states in the tasks of terrorist designation and asset freezing.[13]

Other UN bodies tasked with combating terrorism (including terrorist financing) are the Security Council committee[14] which oversees sanctions measures adopted by the Security Council and its Analytical Support and Sanctions Monitoring Team covering UN sanctions imposed upon the Islamic State in Iraq and Syria (ISIS), al-Qaeda and the Taliban together with associated individuals and entities. UN agencies and entities that have as part of their activities a counter-terrorism role in respect of financial matters include the International Civil Aviation Organization, the International Maritime Organization, the International Organization for Migration and the World Customs Organization, which all work within the CTITF framework. Also relevant here is the Non-Proliferation Committee.[15]

How do we assess these enormous efforts invested by the United Nations in both countering criminal money laundering and terrorist financing? First, as noted above, the United Nations initially approached these problems separately but over time, as the linkages between criminal money laundering and terrorist financing became evident and as the problem of terrorism became more salient, the organisation moved to integrate its approach, as is evident in the widening of UNODC's remit. Secondly, when it came to the issue of terrorist financing, the focus of the United Nations, unlike that of FATF, which rapidly took the global lead in drawing up AML and CFT guidelines, was subordinate to broader political and security concerns. And because a consistent and co-ordinated political approach required consensus on political definitions as to what constituted a terrorist organisation—which was often beyond reach of both the Security Council and the General Assembly— it was always going to be difficult to move outside the established targets of al-Qaeda and later ISIS.

In a 2016 report by the Secretary General to the General Assembly on the implementation of the Global Counter-Terrorism Strategy, it was noted that while the United Nations could co-ordinate policy and support member states, in the final analysis primary responsibility for the strategy's implementation lay with member states. The United Nations itself could, however, point to areas of progress for which it was responsible, and it identified nearly 300 projects in which it was or had been involved, of various shapes and forms, in support of its four pillars, including projects dealing with financial security. Examples here were a UNCCT-led programme on International Good Practices on Addressing and Preventing Kidnapping for Ransom (KFR), which aimed at reducing the ability of terrorist organisations to raise funds through kidnapping especially in West, North and East Africa; UNCCT support of an UNODC-implemented project on Mock Trials on Financing of Terrorism, which sought to enhance the capacity of criminal justice officials in Colombia and Argentina to counter terrorist financing; workshops on combating AML and CFT in Afghanistan; training on strategic trade controls; the creation of a transnational organised crime unit in the Democratic Republic of Congo and the Ivory Coast; workshops on the abuse of non-profit organisations in the Middle East; the promotion of dialogue on the misuse of alternative remittance systems; and strengthening Myanmar's and Kuwait's AML and CFT systems.[16] Education and training are thus central to the United Nations' AML and CFT efforts.

In dealing with the specific problem of financial corruption—a crime that directly links into money laundering—core to the UN efforts is the United Nations Convention against Corruption (UNCAC), which remains the only legally binding global anti-corruption instrument. This convention includes preventive measures that members have an obligation to consider, including declarations of assets of public officials (or what in other contexts are called politically exposed persons), facilitation of the reporting of corruption by public officials, codes of conduct for public officials, transparency in the funding of political parties, the prevention of conflicts of interest within the public sector, and transparent and merit-based employment policies, together with proper remuneration in the public sector.[17] Significantly, UNCAC also defines the supply side—the paying of bribes—as an offence. While, undoubtedly, many member states would benefit from the adoption of such steps, which would act to constrain corruption, bribery and money laundering, implementation by states remains patchy and in many cases nonexistent.[18]

But beyond its achievements, and despite all the well-documented limitations and flaws inherent in the UN process, as the Secretary General Antonio Guterres expressed it, 'the United Nations is the institutional expression of the international community',[19] and so by virtue of involving itself in matters such as criminal money laundering, terrorist financing and corruption, the United Nations reinforces the multilateral importance of these issues and sets goals which members—albeit with varying degrees of commitment—strive to achieve. While because of the definitional problems in respect of terrorism the United Nations has as yet been unable to finalise a comprehensive convention on international terrorism, it has helped develop a normative and legal framework in response to terrorism and terrorist financing by way of its conventions and protocols in respect of human rights, its Global Counter-Terrorism Strategy, and its various relevant General Assembly and Security Council resolutions. Together with the efforts of FATF and regional and national responses, the United Nations provides a global framework and the tools to combat some of the key threats to the security of the global financial system.

The Financial Action Task Force (FATF)

In addition to the UN agencies and bodies there are several multilateral organisations such as the World Bank, the International Monetary Fund and the Egmont Group of International Financial Intelligence Units (FIUs), which have significant AML and CFT functions.[20] But the most important global organisation tasked with defending the integrity of the global financial system and combating key threats is undoubtedly FATF. Thus for students of Finance and Security, it is important to have a grasp of the history of this inter-governmental body, its purpose, the main tools at its disposal (especially its recommendations), and how it has come to dominate the global AML and CFT field.

When developing policies to defend the integrity and stability of the financial system, the most important advantage that FATF has over the UN system is that its primary focus is on finance, a focus that is essentially technical rather than being subordinated to broader political objectives. Such single-mindedness is appropriate to a much smaller and less unwieldy organisation than the UN bodies (with which it co-ordinates) and at least initially, when it was formed, the organisation was governed and directed by like-minded states, which cannot be said about the United Nations when it comes to dealing with terrorism. As a result, while FATF does not involve itself in law enforcement issues, prosecutions and investigations, it is recognised as the essential global driver of legal, regulatory and operational procedures for combating threats to the financial system.

The task force—it is not a formal organisation as its members fund it on a temporary basis—is now over 25 years old. Just as with the United Nations' initial interest in the subject of financial security, FATF originated as a result of the G7 countries' growing concerns about the international criminal drug trade, which was generating billions of dollars in profits with significant social and political costs. It was recognised from the start that only a multilateral global approach would have any chance of being effective in combating the global drugs trade and in protecting the global banking system and other financial institutions from being exploited for purposes of drug money laundering.

The early members were the G7 states (Canada, France, Germany, Italy, Japan, the United Kingdom and the United States), the European Commission and several other countries, and reflected a primarily Western and industrialised worldview, not a surprising fact as the bulk of the global financial sector was located in these jurisdictions. By 2018 FATF consisted of 35 member states and two regional organisations (the European Commission and the Gulf Cooperation Council), representing most of the world's financial centres. In addition there are two FATF observers (Israel and Saudi Arabia), and nine FATF associate members made up of FATF-style regional bodies whose focus is on specific regional AML and CFT issues.[21] Finally there are now various FATF observer organisations, which include the relevant UN agencies we have already mentioned, the Egmont Group of Financial Intelligence Units, the International Monetary Fund, Interpol, the World Bank and many others. FATF's Secretariat was established in 1992 and is located within the Organisation for Economic Co-operation and Development in Paris but remains independent of it. Decisions are made by consensus at the FATF plenary, which meets three times a year.

Within twelve months of its creation, in 1990, FATF had set out its forty recommendations, which established regulatory and legal steps that it recommended members states adopt to prevent the misuse of their financial systems for money-laundering purposes, and to help those states to stop and, failing that, to detect and punish such illegal activity. As FATF correctly notes, 'These measures were the turning point in the fight against money laundering. Up until then, most countries had no legal or regulatory provisions that were specifically targeted at detecting and punishing money laundering.'[22]

Again, like the approach of the United Nations, over the years the goals of FATF have expanded to encompass post-9/11 terrorist financing and then, partly in response to the continuing crisis on the Korean Peninsula, financing of the proliferation of weapons of mass destruction. As threats emerged and mutated, FATF responded accordingly with relevant policy prescriptions—for example, it has recently begun focusing on the threats to the financial system posed by cryptocurrencies. It would not be unfair to say that FATF always seems one step behind the most current challenge, though this is unavoidable, and the task force should rather be judged on the speed of its response and the effectiveness of its proposed policy suggestions.

There are important key roles that FATF undertakes. It identifies and analyses threats to the integrity and stability of the global financial system. It sets AML and CFT standards in the form of its recommendations and related proposals. It also evaluates how effectively member states have implemented those recommendations. Finally, FATF identifies countries that have serious weaknesses in their CFT structures and processes, and engages with them in improving these systems.

Identifying Threats

Typology research stands at the centre of FATF's identification and analytic role. Recognising that threats to the financial sector are constantly altering and related vulnerabilities may themselves ebb and flow, FATF produces detailed research reports into various financial sector vulnerabilities, on the basis of which new guidelines or ultimately recommendations may be generated or the better application of existing recommendations may be suggested.

We have already mentioned a few of these research projects but there now are a substantial number, including professional money laundering (2018), terrorist recruitment (2018), West and Central Africa (2016), emerging terrorist financing threats (2015), gold (2015), the Islamic State (2015), Afghan opiates (2014), non-profit organisations (2014), counterfeiting currency (2013), tobacco trade (2012), human trafficking, trust and company service providers (2010), new payment methods (2010), money remittance and currency exchange providers, free trade zones (2010), securities (2009), football (2009), casinos and gaming (2009), proliferation of nuclear weapons (2008), terrorism (2008), real estate (2007), corporate vehicles (2007), VAT carousel fraud (2007) and trade-based money laundering (2006). In addition, FATF regularly publishes mutual evaluation reports and various country reports.

A review of these studies demonstrates both the breadth and the priorities of FATF's approach to tackling vulnerabilities in the global financial sector. It is in response to these priorities that its guidelines and recommendations are developed.

Recommendations

Underpinning FATF's defence strategy for the global financial system are its forty recommendations. The recommendations consist of guidelines for countries to adopt in their campaigns against money laundering and terrorist financing. Specifically, they include measures by which states can assess where their financial sectors are most at risk. Countries can then allocate resources to the areas where the risks are highest: this is the essence of the risk-based approach whose importance is underlined by the fact that it is found in Part A Recommendation 1 of the current list of FATF recommendations (February 2012).[23] We shall look at this key analytic tool in the following section.

FATF standards consist of the recommendations themselves together with interpretative notes and a glossary. The standards are supported by papers dealing with guidance and best practice which can be used for assisting implementation but do not form part of the standards. In addition to identifying risks, states will, by adopting the recommendations, commit themselves to measures aimed at combating money laundering, terrorist financing and proliferation financing. They will also adopt preventive steps to defend the financial sector, establish powers for law enforcement and supervisory authorities, enhance transparency and beneficial ownership information, and facilitate international co-operation. If properly adopted, the forty recommendations should provide effective means to disrupt attempts to undermine and corrupt the financial sectors of member states, and thereby help in defending the broader integrity of the global financial system.

Over the years the recommendations have gone through several iterations. The first set of recommendations was issued in 1990 and focused on setting guidelines for tackling criminal money laundering related to drug money. By 1996 they had been broadened to combat a wider set of criminal money-laundering threats than that posed by the drugs trade. In 2001, in response to the 9/11 attacks and growing concerns about terrorist exploitation of the global financial system, FATF published eight special recommendations (soon expanded to nine) focusing primarily on steps required to combat the financing of terrorism, which were later integrated into the broader criminal money-laundering set. Other major revisions took place in 2003 and 2012, the latter including a focus on proliferation finance. In this development the FATF approach clearly paralleled that of the United Nations, which similarly moved from a narrow focus on criminal money laundering to a broader set of crimes, to terrorist financing and then other threats as well.

Following an explanation of the risk-based approach in Part A, Part B contains the money-laundering recommendations. It links to relevant UN conventions by calling on states to criminalise money laundering on the basis of the Vienna and Palermo Conventions, and for the seizure and freezing of laundered proceeds and property. Part C then complements the criminal-inspired recommendations with a focus on terrorist financing. Again it links to the UN Terrorist Financing Convention and calls for states to criminalise terrorist financing. States are asked to implement financial sanctions on the basis of targets designated by the Security Council, including Resolutions 1267 (1999) and 1373 (2001). An emphasis is also placed on the need for states to ensure the adequacy of rules dealing with non-profit organisations which may be vulnerable to terrorist exploitation.

Part D deals with what it calls preventive measures. At the state level, governments need to ensure that bank secrecy laws (which may impose privacy and client privilege requirements) do not interfere with the implementation of FATF recommendations. But most of the emphasis here is on the steps that institutions, including financial institutions, need to adopt in order to protect themselves against money laundering, terrorist financing and other threats. This section requires the adoption of adequate measures for customer due diligence so that, for instance, any new customer's identity will be properly verified and the purpose of the business relationship with that client clearly understood. Institutions are also obliged to implement adequate record-keeping procedures in relation to clients and transactions. There are specific due diligence and other requirements in respect of correspondent banking relationships, money or value transfer services, new technologies, wire transfers, and dealings with third parties and higher-risk countries. Any suspicions should be reported to the local financial intelligence unit (FIU). Part D is complemented by Part E, which suggests measures to ensure proper information pertaining to beneficial ownership so that legal entities cannot be exploited for either criminal or terrorist purposes.

Part F deals with supervisory authorities and requires that states ensure that their financial and designated non-financial institutions are subject to effective supervision and regulation, including the implementation of FATF recommendations. There are requirements in respect of the role of financial supervisors, the need to establish FIUs to receive suspicious transaction reports, and the responsibilities and powers of investigative authorities and law enforcement. This section curiously includes a clause on measures to detect cash couriers and the physical transportation of funds and also on the requirement for a set of sanctions against financial and non-financial institutions as well as their directors and managers in the event of money-laundering or terrorist-financing activities.

The final part, Part G, requires states to adopt measures to expedite international co-operation. These include steps to support mutual legal assistance, the freezing and confiscation of laundered proceeds, extradition and other forms of international co-operation that may be required.[24]

Perhaps the key section among FATF's recommendations is Part A dealing with the risk-based approach (RBA), for this recommendation provides the methodology by which states and institutions can dedicate their inevitably limited resources to a wide range of financial threats.

The Risk-Based Approach (RBA)

> *Previous anti-money laundering measures were extremely prescriptive. Solicitors were often left applying the same level of client due diligence to a loyal client selling the family home, as they would for a convicted fraudster who just walked into their office with a suitcase full of cash. Such an approach was hardly an appropriate use of resources and led to a 'tick the box' mentality developing among those in the regulated sector.*

The Law Society[25]

Following the Manchester Arena atrocity in 2017 in which a British Islamist terrorist murdered teenagers attending an Ariana Grande concert, UK security sources let it be known that there were approximately 23,000 individuals in the country who were 'subjects of interest' and who represented a potential threat to national security. It was also revealed that round-the-clock monitoring of all these persons was impossible given the limited resources available to undertake such surveillance, and that ultimately monitoring priorities were determined by an assessment of the perceived seriousness of the individual's intent and also the credibility of the relevant available intelligence.[26]

Since it is beyond the resources of the security services to track all of these suspect individuals, a decision has to be taken as to which 'subjects of interest' to focus on. Who is watched and who is not is essentially a function of some form of risk-based assessment. Sometimes the authorities get it right, sometimes not; and on occasion a terrorist is able to slip though the defensive net created by the security agencies.

This kind of risk-based assessment is also made in the world of financial transactions, especially as governmental authorities and finance and other institutions grapple with the dangers of money laundering, terrorist financing, proliferation finance and other threats to the financial system. It can be no surprise that millions upon millions of transactions are processed daily by banks and other assorted financial organisations. The numbers of deposits and withdrawals, wire transfers, loans and repayments are numerous. The same is true of gatekeeper transactions overseen by lawyers and accountants. These professionals may have hundreds, and in some cases thousands, of clients undertaking numerous transactions all the time. The vast majority of these banking and professional transactions, which constitute the lifeblood of the financial system, are entirely legal, conventional and above board. Only a very small number are indeed illicit, having the aim of circumventing regulations and laws, such as criminal money laundering and terrorist financing.

There are several methodologies the authorities employ to identify money launderers and terrorist financiers amidst a mass of legitimate actors and legal transactions. In theory the most direct approach would be to subject each and every transaction to intense and high-level due diligence. However, the time and cost of such an exercise are potentially prohibitive (if not in many cases impossible) and could deter legitimate business and undermine operating models. Whether such a blanket approach would be successful in helping identify potential wrongdoing is also questionable. An approach treating all transactions equally and subject to the same level of due diligence would inevitably give rise to an even further exponential expansion of the compliance function in banks—a function that has already seen a massive increase in size because of the need to implement a more selective approach to identifying relevant targets.

The challenge is therefore to develop a targeted and responsive mechanism that can act to identify the problem and then direct limited resources at the greatest threats rather than defending the system at each and every point against all manner of pressures independent of the actual dangers they pose. Early on in its existence, FATF rejected approaches based upon the even application of resources where all financial institutions, products and customers received equal attention. The concern was that this would result in a 'tick the box' approach with the aim inevitably focusing on the need to meet regulatory rules rather than mounting a credible defence to defeat a dynamic and protean threat. As far back as 2007, FATF was adamant that efforts undertaken to prevent or mitigate money-laundering and other financing threats should be 'commensurate to the risks identified'. In this way, resources would be best husbanded and allocated most efficiently to meet the threats.

Between 2007 and 2012 FATF refined its approach and in 2012, as part of its revision of the forty recommendations, the organisation produced a definition of RBA that is expressed in Recommendation 1. This recommendation, which is directed at countries, competent authorities and financial institutions, can be divided into three main sections.

The first part sets out the requirement to identify the threat and designate a mechanism responsible for dealing with this threat at the national level: 'Countries should identify, assess, and understand the money laundering and terrorist financing risks for the country, and should take action, including designating an authority or mechanism to coordinate actions to assess risks, and apply resources, aimed at ensuring the risks are mitigated effectively.'[27] The second part of the recommendation calls for the adoption of an RBA to confront the threat:

> Based on that assessment, countries should apply a risk-based approach (RBA) to ensure that measures to prevent or mitigate money laundering and terrorist financing are commensurate with the risks identified. This approach should be an essential foundation to efficient allocation of resources across the anti-money laundering and countering the financing of terrorism (AML/CFT) regime and the implementation of risk based measures throughout the FATF Recommendations. Where countries identify higher risks, they should ensure that their AML/CFT regime adequately addresses such risks. Where countries identify lower risks, they may decide to allow simplified measures for some of the FATF Recommendations under certain conditions.[28]

The third part requires that national authorities ensure that similar measures are adopted at the institutional level: 'Countries should require financial institutions and designated non-financial businesses and professions (DNFBPs) to identify, assess and take effective action to mitigate their money laundering and terrorist financing risks.'[29]

The RBA is, according to FATF, central to the effective implementation of its revised (2012) recommendations and an essential foundation of any country's AML and CFT framework. There is a greater emphasis in the 2012 iteration on both supervision and preventive measures. To this end FATF has provided guidance papers on the application of the RBA to different business sectors such as banking, trust and company service providers, accountants, real estate agents, casinos, dealers in precious metals and stones, money service businesses, lawyers, life assurance and, most recently, virtual currencies. Let us now investigate how the approach works.

At the conceptual level an RBA is grounded upon four key components: threats, vulnerabilities, consequences and risks.

Threats are the starting points for an understanding of criminal money laundering and terrorist financing. They may stem from persons, groups or activities that can cause harm to a state or institution. In the context of criminal money laundering and terrorist financing, these can include criminals, terrorists and their activities.

Vulnerabilities consist of those aspects of the state or the institution that can be exploited by the threats. In the criminal money-laundering and terrorist-financing space they may include weaknesses in the AML and CFT structures or particular services or products that may be particularly attractive to money-laundering or terrorist-financing perpetrators.

Consequences refer to the harmful impact that money laundering or terrorist financing may have on the society or the economy at the national level or, at a lower level, to institutions. Not all forms of money laundering are of equal consequence, and some may present a greater or more immediate danger than others.

Risk is a function of the threat, vulnerability and consequence analysis. It is thus ultimately and essentially a subjective judgement determined by those acting in the relevant department of the state or within an institution. It can be defined as a 'combination of the likelihood of an adverse event (hazard, harm) occurring, and of the potential magnitude of the damage caused'.[30] It forms the basis and starting point of the response.

Thus, in terms of these criteria, both at the national and institutional levels, the relevant authorities and organisations both at the national and institutional levels will produce their own risk assessments based upon the particular threats that face their jurisdiction or organisation and their particular vulnerabilities. Consequences will be determined and a judgement will be reached as to the nature and level of the resulting risk. Proportionate policies and procedures will then be adopted aimed at managing and mitigating the identified risks.

What is meant by 'proportionate' is that where the risks are found to be higher, then enhanced measures should be adopted and, conversely, where risks are low, controls may be reduced. To these ends it should be noted that, as FATF admits, there are no globally accepted methodologies. An acceptable end result will be achieved through the use of a reasonable business judgement applied to identifying and categorising money-laundering risks and setting up proper controls. In theory at least, the potential benefits of adopting such a system will include that of allowing the state or institution to focus on real and present threats rather than those which are less dangerous, while also, at the same time, providing the assessor with a measure of flexibility to adapt to risks that increase or decrease over time.

The effectiveness of risk-based mitigation and prevention measures will be assessed here as part of the mutual evaluation process of national AML and CFT regimes undertaken under FATF auspices.

RBA at the National Level

The conceptual basis of the RBA was set out fully in the FATF 2007 paper on RBA high-level principles.[31] In its 2013 guidance on national RBAs, FATF focused on the three stages of the RBA: identification, analysis and evaluation.[32]

In terms of identification, the risk assessment is initiated by the national authorities through developing a list of potential threats or vulnerabilities (risk factors) which they confront in combating criminal money laundering and terrorist financing. The threat assessment will be based on the national crime threat assessment, typology reports and the collective knowledge of law enforcement agencies. The vulnerability assessment will be grounded upon mutual evaluation reports, supervisors' reports on the regulated sector, and the risk assessments of the various regulated entities.

At the second stage, the likelihood and consequences of the threat will be analysed. The aim is to assign a relative value to the various threats, vulnerabilities and consequences in order to gain an understanding of the risks faced.

The third stage of evaluation involves taking the risks analysed in the previous stage and determining priorities for combating them. These priorities are then inserted into the AML and CFT strategy. Such a strategy will involve prevention or avoidance, mitigation or reduction, or acceptance of a level of risk.

Specifically, a national RBA will be informed by an analysis of the political and legal environment as well as the local economic structure and the scale and scope of the financial services industry. There needs to be a specific assessment of the nature of local payments systems and the relevance of cash-based transactions, as well as a focus on the geographic spread of local financial industry operations and the customer base. An assessment of local criminal activity will be complemented by an assessment of the amounts of illicit money generated domestically and also those generated externally but laundered domestically. An evaluation of the main instruments for laundering monies or financing terrorism will be central to this analysis as will an assessment of the sectors of the legal economy affected by the illegal activities and a determination of the scale and scope of the local underground economy.

In addition to this general assessment, there is a need for a review of the local legal and regulatory framework which underpins the application of the RBA. Concomitantly, a supervisory framework designed to support the application of an RBA should be constructed. Regulators will, in turn, be required to adopt an RBA in supervising financial institutions within their remit. This will be founded on a comprehensive understanding of the types of financial activity undertaken by the institutions in question and the money-laundering and terrorist-financing risks to which these institutions are exposed. National authorities will also need to identify the main stakeholders who will be involved in setting up and implementing an RBA. Key among these would be the various arms of government, financial service regulators, and organisations within both the public and private sectors, including institutions and gate-keepers. It is incumbent upon national authorities to ensure that there is proper and effective exchange of information between these various groups—something it must be said that is easier to specify than it is to achieve.

Table 4.1: Examples of a Risk Evaluation Matrix

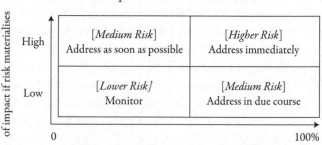

Probability or likelihood that a risk will materialise. Taken from FATF, *National Money Laundering and Terrorist Financing Risk Assessment, 2013*, p. 28.

Examples of such national RBAs include the United Kingdom's *National Risk Assessment of Money Laundering and Terrorist Financing*, first published in 2015 and most recently (at the time of writing) in 2017.

In the 2015 *NRA* the focus was on the domestic risks of money laundering and terrorist financing within the regulated sector. This report dealt with regulated professionals and financial institutions, and the attendant risks connected with cash, new payment methods, and UK legal arrangements and entities. Also within its ambit was the international risk to the United Kingdom's financial sector from money flowing into and out of the country.[33] The 2017 version begins with an assessment of the legal and regulatory framework, predicate crimes and law enforcement as well as the supervisory network. This is followed by a focus on relevant regulated sectors (those sectors with specific AML and CFT obligations), an assessment of the risks associated with the use of cash within the economy, new payment methods, and the international exposure of the UK economy to global money-laundering and terrorist-financing risks. There is a specific chapter on the risks of terrorist financing, reflecting in part the unique characteristics of this particular issue.[34]

In the United States, the *National Money Laundering Risk Assessment 2015* sets out threats in terms of the predicate crimes of fraud, drug trafficking, human smuggling, organised crime and public corruption. The section on vulnerabilities and risks is divided into money-laundering methods using cash, banking, money service businesses, casinos and securities. The United States *National Terrorist Financing Risk Assessment 2015* is a separate document paper. It reviews threats in terms of global sources of terrorist-financing threats facing the United States while vulnerabilities and risks are divided into sections concerning the raising of funds, the moving and placing of funds, and emerging threats.[35]

RBA at the Financial Institutional and Professional Levels

National risk-based assessments are replicated at the institutional level, taking into account the exigencies of private sector operations. An institution obviously has a much narrower and more focused RBA than the state, but the principles are generally the same. As already noted, FATF's 2007 guidance on RBA high-level principles states that at the institutional level: 'a reasonably implemented risk-based process also provides a framework for identifying the degree of potential money laundering risks associated with customers and transactions and allows for an institution to focus on those customers and transactions that potentially pose the greatest risk of money laundering.'[36] Key to the approach at the institutional level is the identification of so-called risk categories: country and geographic risk, customer risk and product or services risk. These categories encapsulate the key money-laundering and terrorist-financing risks facing financial institutions daily and serve to guide the institutions in the direction of implementing proportional measures and controls to contain such threats.[37]

There can be no generally accepted means of judging at the state or institutional level whether a particular state or region represents a higher risk than others. Each institution needs to prepare its own such country risk analysis, which will be informed by whether the country is subject to sanctions and embargoes. Should the country be identified by 'credible sources'[38] as having

AML and CFT deficiencies, providing funding or other support for terrorist activities, or exhibiting high levels of criminal activity or corruption, then this information in conjunction with other risk factors will require added diligence and counter-measures. Obviously, proposed transactions involving countries such as Iran, North Korea and Russia should raise alarm bells, but there are also numerous grey areas where further analysis will be required.

Customer risk involves identifying money-laundering and terrorist-financing threats posed by a particular customer or category of customer, offset by any relevant mitigating factors. Red flags here could include customers conducting unusual transactions such as frequent and unexplained movement of funds between institutions in varying geographic locations, unexplained geographic distances between the customer and the institutions, as well as frequent and unexplained movements of accounts to different institutions. In addition, institutions will need to pay particular attention to customer relationships where the entities involved have unclear beneficial ownerships, cash-intensive businesses such as currency exchanges, money transfer agents and gambling activities, charities and other non-profit organisations, especially those operating across borders, gatekeepers such as accountants and lawyers acting on behalf of suspect clients, and of course politically exposed persons (PEPs).

A financial institution's risk assessment is also informed by a consideration of the particular products and services it offers, including international correspondent relations, especially when these services are provided for non-customers who have not been subjected to proper due diligence. Attention needs to be given to the provision of international private banking services and services involving banknotes and also precious metal trading and delivery. Also relevant here are services that cross international borders or provide anonymity such as global wire transfers, private trusts and investment companies, stored value cards and online banking services.

When higher-risk situations are identified, financial institutions should implement a range of controls to mitigate the threat of money laundering and terrorist financing. These can include, in an augmented form:

- sensitivity in respect of higher-risk transactions and customers;
- 'know your customer' checks and enhanced due diligence;
- levels of controls and added frequency of reviews of banking relationships; and
- escalation for the approval of new relationships and accounts.

In addition to the country, customer and product categories, the RBA highlights other markers that may be useful for financial institutions in identifying money laundering and terrorist financing. These can include an assessment of the purpose of an account or a relationship (for instance, an account opened to allow significant cash flows from a previously unknown commercial organisation should attract enhanced due diligence): the deposit of a large amount of assets or an unusually big transaction from a low-risk-profile customer (these may be viewed as a higher risk); the duration of a relationship and the regularity of that relationship (short-standing infrequent relationships should pose a greater risk); familiarity with the country of the client including its regulatory regime (lack of knowledge should add to the risk); the use of intermediate corporate vehicles and other structures that do not appear to have a clear commercial rationale, are unduly complex or lack transparency (these will ultimately increase the risk).[39]

Nevertheless, despite the implementation of an RBA-based system either at the national or institutional levels, the approach cannot be expected to identify and detect all instances of money laundering or terrorist financing. The system is by no means impregnable. What it can do is increase the difficulties facing money launderers and terrorist financiers as they seek to exploit the system; it may also work to deter them and even help apprehend them. The implementation of such a system is not without costs to the implementing financial institutions, in terms of direct expenses and lost business.

Application of the RBA to the Banking Sector

The formal banking sector remains one of the key sectors vulnerable to criminal money laundering, a vulnerability that is also true in respect of terrorist financing, at least as concerns the retail sector. Thus FATF's Recommendation 26 requires states to ensure that banks are subject to adequate AML and CFT regulations while Interpretative Note 26 requires supervisors to direct supervisory resources to areas of highest risk.[40]

The FATF's RBA Guidance for the banking sector encourages supervisors to develop an understanding of criminal money-laundering and terrorist-financing risk and to work to mitigate that risk. It sets out a general approach to the supervision of banks in terms of RBA and also provides guidance for banks in respect of the RBA. As for the banks' responsibilities, they are expected to undertake a risk assessment in terms of the nature of their business, their target markets, the number of their customers identified as high risk, the jurisdictions to which that bank is exposed, and the volume and size of its transactions in the context of the typical activity of the bank and its customer profile.

The Guidance helpfully outlines the types of criminal money-laundering and terrorist-financing risks it views as associated with various kinds of banking activity. So, for example, risks associated with investment banking include layering and integration and the transfer of assets between various parties for cash and other assets. Wealth management risks are associated with confidentiality, banking secrecy, PEPs and the identification of beneficial ownership. Retail banking risks are linked to the volume of transactions and the provision of services to cash-intensive businesses. And correspondent banking relations are viewed as by their very nature vulnerable to penetration especially when dealing with jurisdictions with lax AML and CFT structures.[41]

This analysis is then followed by a focus on the requirement for banks to mitigate risks through identification and verification of the purpose and intended nature of business relationships, with an emphasis on due diligence and enhanced due diligence. There is thus a requirement for ongoing customer due diligence and monitoring, and therefore not just when customers are on-boarded. In addition, pursuant to Recommendation 20, a bank will need to report any suspicious transactions to its national FIU. Finally, there is also a section on banks ensuring and monitoring compliance, including in respect of vetting and recruitment, training of staff and internal controls.[42]

Let us now turn briefly to some examples of the implementation of the RBA to the banking sector by national supervisory authorities.

In the United Kingdom the Financial Conduct Authority (FCA), the conduct regulator for approximately 56,000 financial services firms and financial markets in the country and the prudential regulator for more than 18,000 of those firms, classifies firms subject to relevant AML legislation according to the money-laundering risk they may present. Such a classification is based upon an assessment by the FCA of the type of business, the types of products and services offered, and the jurisdictions in which the bank operates. The level of categorisation determines the FCA's supervisory system.

The highest bands are dealt with by the Systematic AML Programme (SALMP), which can last several months and operates on a four-year rolling cycle. The FCA will interview relevant staff including senior management and the compliance department. It will also rigorously test the firm's AML and CFT system. Second-band firms are subject to two- to three-day visits, which occur every two years. The lower-risk banks are visited depending upon the occurrence of relevant events or are subject to thematic reviews consisting of off-site assessment of relevant policies.[43]

The FCA also publishes non-binding guidance, which sets out what it expects from a firm's AML and CFT system as well as examples of good and bad practice. By answering questions contained in the FCA's guidance, firms can test the efficacy of their systems.[44]

In the United States the Federal Banking Agencies (FBAs) provide the supervisory function for their regulated financial institutions. They determine whether depository institutions are fulfilling the obligations of AML and the Bank Secrecy Act (BSA) 1970, including making suspicious transaction reports to the relevant enforcement agencies.

Key to consistent application of BSA, AML and CFT procedures, the FBAs follow the procedures set out in a manual—the *Bank Secrecy Act/Anti-Money Laundering Examination Manual* produced by the Federal Financial Institutions Examination Council—which provides guidelines to institutions for meeting their AML and BSA requirements and FBA supervisory expectations. The manual incorporates information from the US *National Money Laundering Risk Assessment* and various other key documents, and identifies services, products, entities and persons that constitute risks for the financial sector.

Reviews of depository institutions are conducted on a twelve- to eighteen-month cycle by way of a mixture of on-site and off-site reviews, though for larger institutions some FBAs may retain on-site examiners who provide constant supervision. The FBAs work closely with the US's financial intelligence unit the Financial Crimes Enforcement Network (FinCEN).[45]

Measuring Effectiveness: Iceland as a Case Study

Has FATF accomplished any of its main objectives? There are two issues here. One is the extent to which FATF has secured the global implementation of its recommendations; the second is how effective these recommendations have been in countering money laundering, terrorist financing and FATF's other stated goals.

Certainly some of the recommendations were rapidly adopted by national authorities. By 1993 nearly all FATF members had criminalised money laundering and by the following year all member states permitted their banks to report suspicious transactions to relevant authorities. By 2004, the FATF-style regional bodies as well as the World Bank and the International Monetary

Fund agreed to employ the same methodology in assessing compliance with the FATF recommendations both in terms of AML and CFT.

But it is one thing to set goals in the form of recommendations to which national authorities must work. It is another for those authorities to effectively adopt these recommendations and then maintain them properly. Further, given the interconnectedness of the global financial system, any defensive system will, as FATF notes, inevitably be as strong as its weakest link. FATF itself of course has no enforcement authority and relies on the voluntary actions of its member jurisdictions. Not all states have demonstrated equal commitment or capacity to fully implement all of FATF's recommendations.

In its early years FATF members self-reported as to their AML systems, but over time this method was supplanted by a peer-reviewed mutual evaluation process whereby a jurisdiction's AML and CFT measures, its legislation, law enforcement and financial intelligence are assessed by other members of FATF. This evaluation results in two ratings. First, there is a technical compliance rating system that reflects how well a country has implemented technical requirements linked to FATF's 40 recommendations. These technical measures are ranked as compliant, largely compliant, partially compliant or non-compliant. Secondly, there is the effectiveness rating system that ranks the actual effectiveness of a country's AML, CFT and counter proliferation finance mechanisms. This latter system works by identifying a set of eleven 'Immediate Outcomes' which are measured on a spectrum ranging from a high level to a low level of effectiveness.

These immediate outcomes are identified as the following:

(i) There is an understanding of risks to the financial systems and there exist co-ordinated AML, CFT and counter-proliferation measures.

(ii) There are appropriate levels of international co-operation.

(iii) There is an appropriate level of relevant supervision.

(iv) Financial and other institutions should apply AML and CFT measures commensurate with the risks they face, and report suspicious transactions.

(v) Adequate information on beneficial ownership is made available, and legal arrangements cannot be exploited for money-laundering and terrorist-financing purposes.

(vi) Financial intelligence is properly employed.

(vii) Money-laundering offences are investigated and prosecuted.

(viii) Proceeds of crime are confiscated.

(ix) Terrorist-financing offences are investigated and the relevant persons prosecuted.

(x) Terrorists are prevented from raising and moving funds and from abusing non-profit organisations.

(xi) Proliferators are prevented from raising and moving funds.

Immediate Outcomes then feed into mid-range Intermediate Outcomes which represent so-called 'thematic goals' of an AML/CFT system that FATF considers will protect the financial system and contribute to broader security. These involve (i) policy co-ordination and co-operation aimed at mitigating money laundering and terrorist financing risks; (ii) proceeds of crime and funds in support of terrorism being prevented from entering the financial and other sectors (or being detected and reported by those sectors); and (iii) money laundering threats being detected and disrupted, criminals being deprived of their illegal proceeds, terrorist finance threats being detected and disrupted, and terrorists being separated from their resources and sanctioned.

Immediate Outcomes and Intermediate Outcomes are ultimately dedicated to support of the 'High Level Objective' which is defined as when 'Financial systems and the broader economy are protected from the threats of money laundering and the financing of terrorism and proliferation, thereby strengthening financial sector integrity and contributing to safety and security'.[46]

As FATF itself notes, it is this mutual evaluation that gives the FATF its teeth. A follow-up process is undertaken to ensure that countries with weaknesses identified in the evaluation take steps to address them.

Take, for example, the Mutual Evaluation Report on Iceland published in April 2018, which gives a flavour of the nature of the evaluation process. Key findings of the nearly 200-page document point to the existing deficiencies in the Icelandic approach and make recommendations for the way forward. While Iceland was complimented on taking initial steps to understand the money-laundering and terrorist-financing threats it faces on the basis of its first national risk assessment, that assessment was regarded as too theoretical and not founded upon actual vulnerabilities in the Icelandic financial sector. Nor, it was noted, had the assessment been co-ordinated with previous risk assessments undertaken by the Icelandic police. In addition, the lack of domestic co-ordination between relevant authorities 'negatively affects Iceland's entire AML and CFT regime'. The view was that until then Iceland had not perceived money laundering as a priority and there was a lack of resources dedicated to supporting AML. While the use of financial intelligence had improved, the proper employment of this intelligence was hampered by limited suspicious transaction reporting, and lack of sharing of information on cross-border movements of currency, non-profit organisations, and beneficial ownership. There was lack of consideration of terrorist-financing vulnerabilities, and limited investigative expertise served to restrict Iceland's counter-terrorist financing policy. An understanding of the dangers of money laundering and terrorist financing existed among the larger Icelandic commercial banks, but this was not replicated among the smaller financial or non-financial institutions. Significantly, it was pointed out that supervisory measures were not yet being conducted using a comprehensive risk-based approach. More work also needed to be done by the Icelandic authorities on beneficial ownership information.

It should be noted that Iceland had undergone a FATF assessment in 2006. Yet, twelve years later the abovementioned deficiencies were still being identified. It seems that Iceland had made only limited progress in upgrading its AML and CFT capabilities. In the 2018 evaluation FATF did accept that Iceland's regime had undergone significant reforms since 2006, especially in identifying supervisory limitations in respect of money services providers. There had also been changes in relation to correspondent banking. Overall, Iceland's technical compliance framework was strong, especially in terms of law enforcement powers and international co-operation, but there remained weaknesses, including in respect of transparency of legal structures and out-reach to non-profit organisations. In summary: 'In terms of effectiveness, Iceland achieves substantial results in international co-operation and moderate results in terms of collection and use of financial intelligence, investigation and prosecution of [money laundering] and confiscation of assets and instrumentalities of crime. More significant improvements are needed in other areas.'[47]

As this report shows, the reviews can be unsparing and in part highly critical. Nevertheless, despite the lapse of over a decade since Iceland's previous evaluation, and despite clearly significant progress in some areas, there remained room for improvement in the implementation of FATF recommendations. Formal commitment to FATF therefore does not necessarily equate to immediate implementation of all its standards.

Non-Cooperative Jurisdictions and High-Risk Jurisdictions

While according to the Icelandic case study, there may be lacunae in the defensive system created, there is little doubt that the country has put in place robust structures and processes to combat threats to its financial system. Furthermore, few question the seriousness of the Icelandic authorities' commitment to working towards enhancing their AML and CFT structures and processes. That cannot be said about all jurisdictions.

In 2007 FATF created the International Co-operation Review Group, which works with identified non-cooperative countries and territories to improve seriously deficient AML and CFT systems. FATF believes that a naming and shaming policy is an effective means to get wayward jurisdictions on board with the FATF recommendations. FATF has thus over the years publicly identified 65 countries as deficient. However, more than 50 of these have made the required reforms to their AML and CFT processes and structures and are no longer subject to FATF monitoring. As of early 2018, North Korea, Ethiopia, Iran, Iraq, Serbia, Sri Lanka, Syria, Trinidad and Tobago, Tunisia, Vanuatu and Yemen remained identified as high-risk or otherwise monitored jurisdictions.[48]

In June 2018, FATF identified Pakistan as a country with deficiencies in its AML and CFT systems, which required remediation, while also noting that Iran had not implemented an agreed action plan to address its problems in this area. On the other hand, both Iraq and Vanuatu had made good progress in addressing their relevant deficiencies and would no longer be subject to FATF monitoring under its AML–CFT compliance process.[49]

While non-compliant jurisdictions may remain an AML and CFT risk and while even among FATF member states compliance levels may differ, if one adds the number of FATF members to the members of the regional bodies (the so-called Global Network) there are a significant number of states committed in varying degrees to the standards embodied within the FATF recommendations. FATF aims to steadily narrow the space where criminals and terrorists can act, raising the costs of their activities and thereby defending the financial sector against malevolent infiltration.

FATF and the Combating of Bribery and Corruption

In Chapter 1 we noted the threat to security in general and financial security in particular posed by the dangers of corruption and bribery. FATF views bribery and corruption as inextricably linked to money laundering, and regards AML and CFT as powerful tools in combating these challenges. Consequently, corruption issues were first specifically tackled during FATF's third-round mutual evaluation process which assessed members' compliance with FATF's 2003 version of the recommendations. This analysis involved determining whether states could demonstrate that they had implemented measures to prevent corruption through transparency, the adoption of good governance principles, proper ethical and professional requirements, and reliance on a court system which could ensure that judicial decisions targeting corruption were properly enforced. The revised 2012 recommendations and the fourth round of mutual evaluations together with the G20 leaders' declaration at the 2013 St Petersburg Summit reinforced this anti-corruption focus.[50]

While FATF recommendations focus on AML and CFT, they also cover corruption and bribery risks by requiring members to make corruption and bribery both predicate offences for money laundering and by specifically requiring financial entities to mitigate PEPs risks and establish mechanisms to recover through confiscation the proceeds of such crimes. FATF also encourages AML, CFT and anti-corruption officials to increase co-operation and co-ordination.[51]

In combating corruption FATF places much emphasis on preventive measures and record-keeping. Customer due diligence, especially in respect of PEPs, is emphasised. FATF admits that 'effective implementation of the PEPs requirements has proven to be challenging for competent authorities, financial institutions and [non-financial institutions]', but stresses that appropriate actions in this regard definitely increase the chances of detecting corruption. FATF notes that PEPs typically, following a crime, attempt to rapidly move their money out of their home jurisdiction so as to avoid detection. Wire transfers and cross-border transportation of cash and bearer negotiable instruments are favoured PEP techniques and must be subjected to oversight. Efforts should be made to increase the transparency of legal arrangements and to enhance the detection, investigation, prosecution and confiscation of the proceeds of corruption and bribery. Central to these efforts, as usual, is enhanced international co-operation and engagement with the private sector. As FATF explicitly emphasises:

> Regulatory authorities and FIUs should work in partnership with the private sector to enhance the implementation of AML/CFT measures that combat corruption, including PEPs requirements. Open and constructive lines of communication between regulatory authorities and financial institutions are important to ensure that PEPs requirements are well understood and implemented effectively. In this respect, regulatory authorities should provide support to financial institutions and DNFBs [designated non-financial businesses] regarding the applicable measures on PEPs through the issuance of guidance

or detailed instructions which clearly outlines their legal and regulatory requirements.[52]

Whether the private sector is in fact receiving adequate and timely actionable intelligence on PEPs is sometimes a matter of real contention.

FATF thus continues to evolve to meet the shifting challenges posed by threats to the integrity of the global financial system. Following the 2016 terrorist attacks in Paris, the organisation adopted a new counter-terrorist operational plan focusing on four main areas:

- a renewed effort to better understand terrorist-financing risks;
- a push towards enhanced information sharing;
- enhanced efforts towards increasing engagement with the criminal justice system with the aim of leading to more successful terrorist-financing investigations, prosecutions and convictions; and
- delivering better co-ordination with FATF's regional bodies.[53]

Yet, while FATF drives forward the global AML and CFT agenda, it is at the national level that FATF recommendations are implemented and enforced.

5

NATIONAL ANTI-MONEY LAUNDERING
AND COUNTER-TERRORIST FINANCING
ARCHITECTURES

*Stopping the flow of money to terrorists may also be one of the very best ways we
have to stopping terror altogether.*

United States Treasury General Counsel David Aufhauser, 2003[1]

Mr Shah Takes on HSBC

By September 2006 HSBC was becoming increasingly suspicious about the
activities of one of its clients, Jayesh Shah. Mr Shah was a wealthy businessman
who conducted business in several central African countries as well as in
Zimbabwe where he lived. Three months earlier Shah had transferred some
$28 million to HSBC in London from an account he held with Crédit
Agricole Indosuez (Suisse) SA in Geneva.

He explained to HSBC that the reason for this transfer was that his
account at Crédit Agricole had become vulnerable and that someone was
trying to gain access to it and make unauthorised withdrawals. The money was
successfully moved to the HSBC account but shortly thereafter Mr Shah
began instructing HSBC to return the funds to Crédit Agricole. Much to
Shah's annoyance, HSBC began prevaricating. The reason for the bank's tardi-
ness was that HSBC suspected that the funds in the account were in fact
criminal property and as a result the bank ultimately filed five suspicious activ-

ity reports to the Serious Organised Crime Agency (SOCA) in London, outlining their concerns and requesting instructions as to how to proceed. While they awaited SOCA's instructions, the bank refused to act on many of Shah's requests. While they sought to avoid 'tipping him off' by setting out to him their real concerns (and thereby potentially allowing him to take evasive action), they explained to Shah that the cause of the delays was that the bank needed to comply with various statutory requirements.

Unable to pay his creditors, Shah soon ran into problems with a former employee based in Zimbabwe to whom he owed a substantial amount of money. In response, the disgruntled man alerted the Zimbabwean authorities that Shah's UK accounts had been frozen and that Shah was suspected of money laundering. The Zimbabwean authorities moved rapidly, freezing Shah's accounts and confiscating his investments in the country. Furious and alarmed, Shah went on to sue HSBC for his losses, which he claimed exceeded $300 million dollars.

The legal reasoning behind the court's decision does not concern us here. What we should note is that after four and a half years, three visits to the Court of Appeal and 27 days of trial, the High Court of Justice dismissed Shah's claim and by so doing confirmed a bank's right to delay execution of a customer's instruction and also refuse to provide information to its client where the bank has a suspicion of money laundering and has properly notified the relevant authorities as to those suspicions.[2]

When it came to dealing with money laundering, HSBC had followed the procedures established by law. It had identified a suspicious transaction and had reported it accordingly, and as a result was able to defend itself against a very large claim for damages alleged by the plaintiff to have been caused by the bank's anti-money laundering (AML) actions.

Students of Finance and Security are thus well advised to develop an understanding of not only global AML and counter-terrorist financing (CFT) regimes, such as that of the Financial Action Task Force (FATF), which set general policy guidelines, but also key national AML and CFT approaches. It is of course at the national level where the front line of the war against criminal money laundering and terrorist financing is to be located. What is presented here is not a full legal or historical-legal analysis but rather a very general, high-level overview of some of the main national laws, structures and processes that will enable a student of Finance and Security to appreciate the main elements of some of the most important national AML and CFT regimes and how these ultimately impact upon decision-making at the level of business.

National anti-money laundering and counter-terrorist financing architecture refers to the combination of laws, procedures, structures and participants that together serve to provide the means by which a state can identify illegal activity, work to curtail that activity, gain control of the illegal proceeds that derive from such activity, and prosecute those who engage in criminal and terrorist finance activities which ultimately endanger financial and national security.

Because this architecture is grounded in laws and practices that are local and unique, no two national systems are totally the same. It is for this reason that a considerable amount of United Nations and FATF activity is dedicated towards co-ordinating and harmonising national AML and CFT responses. Consequently, it is necessary to be selective in deciding upon which national regimes to focus as a broader comparative analysis would inevitably sacrifice necessary depth for breadth.

Given the centrality of the City of London's financial sector both to the United Kingdom's economy and the global economy, British authorities have developed one of the most complex and sophisticated set of laws and proce-dures to combat criminal money laundering and terrorist finance. Indeed, as the FATF 2018 Mutual Evaluation Report of the United Kingdom noted, 'The United Kingdom has implemented an AML/CFT system that is effec-tive in many respects.'[3] The United Kingdom has a relatively wide definition of money-laundering offences and a person may be found guilty of contraven-ing such laws by possession of property and the suspicion that such property is in fact the proceeds of a crime. In addition, the country has an exacting AML and CFT regime impacting on a regulated sector covering, inter alia, financial institutions and professional finance and related gatekeepers. For those in this regulated sector subject to AML and CFT requirements there are wide-ranging reporting requirements and significant sanctions for breaches. Central to the country's approach is a coalition between the public and private sectors. Furthermore, because of various historical connections and political and common law linkages the United Kingdom's approach to AML and CFT has been influential in the development of that of other juris-dictions. For students seeking to apply an AML and CFT model to their own jurisdictions, the United Kingdom provides a useful heuristic tool and—though not in all cases—a transferable template or at least a model against which it is possible to assess various other systems.

Attention must of necessity also be directed to the United States' AML and CFT system. The global significance of the US financial sector, the size of its AML and CFT network, the importance of American policies in determining global approaches, the implications of its extraterritoriality, and the huge penalties that await those that contravene American laws make a general appreciation of the relevant American AML and CFT architectures critical to all students and practitioners in the field of Finance and Security. Core to the US approach in combating criminal money launderers and terrorist financiers are also a coalition between public and private interests and an ability to reach beyond American borders in enforcing AML and CFT policies.

Significant United Kingdom AML and CFT Legislation

The United Kingdom was a founding member of FATF, whose recommendations provide significant input for European Union Anti-Money Laundering Directives. As of the autumn 2018 there have been five of these directives, which in turn provide (or will provide) the source for a significant amount of current UK AML and CFT legislation. That being said, in some respects the United Kingdom's approach in countering criminal money laundering and terrorist financing remains ahead of its European counterparts.[4]

The key piece of primary legislation in the area of UK AML is the Proceeds of Crime Act 2002 (POCA 2002).[5] For purposes of combating terrorist financing, the most significant pieces of legislation are the Terrorism Act 2000 (TA 2000), the Anti-Terrorism Crime and Security Act 2001 (ATCSA 2001), the Terrorist Asset-Freezing etc. Act 2010 (TAFA 2010) and, most recently, the Sanctions and Anti-Money Laundering Act 2018 (SAMLA 2018), which is relevant to money laundering, terrorist financing and sanctions. The key secondary legislation is the Money Laundering, Terrorist Financing and Transfer of Funds (Information on the Payer) Regulations 2017 (Money Laundering Regulations) which is also relevant to both criminal money laundering and terrorist financing.[6] There are of course other pieces of critical legislation which together form a rich and detailed national AML/CFT tapestry.[7]

Part 7 of POCA 2002 establishes three types of money-laundering offences relating to the proceeds of crime:[8]

1. Concealing—concealing, disguising, converting or transferring the proceeds of crime, or removing these proceeds from the jurisdiction of England and Wales—the main money-laundering offence (s. 327);
2. Arranging—entering into or becoming part of an arrangement in which the person knows of or suspects the retention, use or control of crime proceeds—the aiding or abetting offence (s. 328); and
3. Handling—acquiring, using or possessing the proceeds of crime—the handling of stolen goods offence (s. 329).

All these elements require some knowledge or suspicion of money laundering, and potentially a whole range of people that may be involved in the laundering are covered by these offences. These include a person handling the proceeds of a fraud or robbery, the person moving the funds such as the lawyer, accountant, courier or professional money launderer, and the banker receiving the funds. POCA 2002 will also catch not only the criminals undertaking the laundering but potentially those benefiting from it—for example, those residing in property or living the high life paid for by laundered funds. Criminal property represents a person's benefit from criminal conduct and includes not only money but goods as well or any profits generated as a consequence of the original offence. If the offence is committed abroad—if, for example, the criminal activity takes place in Russia, Mexico or Australia—any benefits derived from that activity will be construed as criminal property, provided that the activity was a criminal offence in both the United Kingdom and the overseas country when it took place. That the requirement is for the offence to be illegal both in the United Kingdom and abroad deals, inter alia, with the so-called Spanish Bullfighter problem, whereby, had it not been for this requirement, a bank in the UK in receipt of a matador's salary would be committing an offence under s. 327. There are various exceptions to this last rule, one of these being terrorism.

POCA 2002 applies to both people and organisations though the latter will usually not be liable unless guilt can be attributed to an individual.

Presumably, most compliance practitioners, finance professionals and gatekeepers will not knowingly involve themselves in laundering, or abetting or handling illegal proceeds. More pertinent to them is a set of offences under POCA 2002 linked to these activities and that relate to the failure to report money laundering to the relevant authorities. These offences pertain to a person's action or lack thereof upon their discovery of potential money laundering:

1. Failing to disclose—a person working in the 'regulated sector'[9] knows or suspects, or has reasonable grounds for doing so, that another person is engaged in the offences of laundering, helping and handling offences located in ss. 327-9, and fails to report his knowledge or suspicion to a relevant officer (s. 330).[10]
2. Failure to disclose—the failure of the relevant nominated officer who is tasked with receiving such disclosures under s. 330—unless there is a reasonable excuse—to use this information to make further disclosures as required (s. 338).

3. Tipping off—a person working in the regulated sector who knows or suspects that someone is under, or may soon be under, investigation for money laundering may tell that party and thereby prejudice any investigation (s. 333A). This is actually more difficult to avoid than it sounds as the suspected money launderer may have had his account frozen or his transaction halted and may be peppering the bank with queries as to what is happening with his money and why the bank is obfuscating in relation to his instructions. We encountered an example of this in the Shah case.

4. Prejudicing the investigation—here a person knows or suspects a money-laundering investigation is or will take place and makes a material disclosure to any other person that is prejudicial to the investigation. It also applies to tampering with relevant material (s. 342). This particular offence covers both regulated and non-regulated sectors.

The maximum sentence for money laundering is 14 years imprisonment.

Many of the provisions of POCA 2002 apply to acts of criminality and terrorism alike, but counter-terrorism efforts are reinforced with their own set of legislative instruments. As the British lawyer Jonathan Peddie forcefully argues:

> counter-terrorism legislation is a complex patchwork created by Parliament over a period characterised by extraordinary volatility; in the diversity of threats to national security; in governments' responses to those threats; in turnover of the governments themselves; and in the unprecedented backdrop of the global financial crisis, which itself has created increased economic fragility on a global and national scale and has further strengthened the interconnection between economic stability and terrorist activity.[11]

The British authorities, spurred on by 9/11 and related attacks, the emergence of home-grown jihadist terror and residual Northern Ireland terrorism, recognised that combating terrorist financing was vital in protecting the country's security. Countering terrorist financing thus forms a key part of the CONTEST[12] counter-terrorism strategy, with the aim being to reduce the terrorist threat to the United Kingdom and its interests overseas by depriving terrorists and violent extremists of the financial resources and systems required for terrorist-related activity.

TA 2000 defines terrorist property in Part III of the legislation as money or property that is likely to be used for terrorist purposes, including the resources of proscribed organisations, proceeds resulting from the commission of terrorist acts, and also the proceeds of acts carried out for terrorist purposes.

The key terrorist-financing offences are set out in sections 15–18 and include raising, providing, receiving, using or possessing funds or other types of property that the offender has reasonable cause to suspect may be used for the purposes of terrorism, including benefiting proscribed organisations; and becoming part of an arrangement that involves making funds available for the purpose of terror or that abets the retention or control by another of terrorist property.

Within TA 2000 there are provisions that allow the police to request from financial institutions certain types of financial information concerning their clients, and also measures whereby accounts may be monitored for purposes of a terrorist investigation. Following conviction of a person for terrorism finance offences, courts may make forfeiture orders whereby money or other property in the possession or under the control of the convicted person may be seized. While investigations or criminal proceedings are still pending, a restraint order may be issued. This has the effect of restricting a person with notice of the order from removing from the country any property in relation to which a forfeiture order has or could be made.[13]

ATCSA 2001, which came into force soon after the 9/11 terrorist attacks, contains provisions that deal with freezing orders and seizure, retention and forfeiture of suspected terrorist funds. Significantly, if HM Treasury considers that a foreign government or residents of a foreign country threatens persons in the United Kingdom or the nation's economy, then it may make an order freezing terrorist funds.[14] The provisions work by prohibiting persons subject to British law (and this will include banks and other financial institutions) from making funds available to those individuals identified in the order. This has only been employed twice, once in 2016 when freezing the funds of two Russian nationals who had been found by an independent inquiry to have acted on the authorisation of representatives of Russia and used polonium-210 administered into a cup of tea to poison a former Russian intelligence agent, Alexander Litvinenko, in a London restaurant.[15] The other case occurred during the 2008 financial crisis when the funds of an Icelandic bank, Landsbanki Islands hf, were frozen. The fact that anti-terrorism legislation was being used against an Icelandic bank caused great offence in that country.

In August 2007 three brothers and British citizens, A, K and M, were told by HM Treasury that there were grounds for suspecting them of supporting acts of terrorism as defined by Article 4 of the Terrorism (United Nations Measures) Order 2006 (TO 2006). In other correspondence A and M had been identified as facilitators of al-Qaeda, and there was also evidence to suggest that K and M had gone to Pakistan for the purposes of not only engaging in terrorist training but also of delivering money to al-Qaeda. Another person, G, was informed that he too had been designated under TO 2006 and that the Sanctions Committee of the Security Council of the United Nations (the so-called 1267 Committee) had added his name to its consolidated list of individuals and entities subject to sanctions.[16] As a result, all his economic resources were frozen but no mention was made at the time of what domestic law he was being punished under. G was later informed that he was a designated person under UN al-Qaeda and Taliban sanctions orders.[17] A fifth person was added to the UN consolidated list, making him also subject to various financial sanctions. The Supreme Court in reviewing this decision noted that the various orders were unlawful as they lay outside the ambit of British parliamentary scrutiny.[18] As a result, the government rapidly enacted TAFA 2010.

TAFA 2010 provides powers to HM Treasury to indefinitely freeze assets and also to compel organisations in the private sector to provide required information. It gives national effect to UN Security Council Resolution 1373 (2001), which is also concerned with with the financing of terrorism, and also EU Regulation 2580/2001, which deals with the freezing of assets of persons designated within the European Union. HM Treasury has the power to make interim and final designations of a person who was or may have been involved in terrorist activity or an institution owned or controlled by that person or acting on behalf of or at the direction of such a person. At the same time, HM Treasury could request information and documents from designated persons and from institutions such as banks which may have had such information. The aim is to monitor compliance with or detect evasion of Part 1 of TAFA 2010 dealing with terrorist asset freezing. Concomitantly, relevant institutions were also required to report designated persons if they know or have reasonable suspicion that such persons are committing relevant offences.[19]

British AML laws derive, as we have noted, from FATF and EU legislation. SAMLA 2018, which as the name suggests, has sanctions as well as anti-money laundering elements, enables UK authorities to put in place their own post-Brexit AML/CFT regime. (We will deal with the sanctions aspects in Chapter 6.) Significantly, it makes it easier to freeze terrorist assets (on a 'reasonable grounds to suspect' threshold and not on a 'reasonable belief' plus 'necessity' threshold). SAMLA 2018 also empowers the Secretary of State to provide for regulations which support the detection, investigation or prevention of money laundering and terrorist financing as well as implementing relevant FATF standards. While SAMLA 2018 provides powers to pass secondary AML and CFT regulations, it does not impose new AML obligations. The new regulations cannot revise substantial criminal offences under POCA 2002, which must be done by way of primary legislation. What constitutes terrorist financing remains that which is defined in the relevant sections of the TA 2000. It will of course be a criminal offence not to comply with future regulations.

It is possible that over the next few years, UK and EU AML approaches will diverge, but given the centrality of FATF requirements, divergences should be held in check at least to some extent. In the meantime, the United Kingdom co-operates with the European Union in terms of extradition and expedites processes in relation to the transfer of evidence for criminal investigations and asset freezing. Whether this will long survive Brexit remains to be seen.

The sets of legislation discussed above are buttressed by secondary legislation—specifically the Money Laundering Regulations, which are relevant to both criminal money laundering and terrorist financing and in many ways act as a linchpin between the primary legislation and those conducting regulated business and transactional work in the United Kingdom.[20]

In order to avoid committing an offence under the Money Laundering Regulations, businesses are required to undertake a set of AML and CFT measures; and the supervisory authorities monitor that they do so. This is the glue that holds the public–private AML and CFT coalition together. The key influences of the FATF risk-based approach discussed in Chapter 4 can clearly be seen here. Firms should, among other steps:

- Conduct a written money-laundering and terrorist-financing risk assessment;
- Implement systems, policies, controls and procedures aimed at tackling money-laundering and terrorist-financing risks and thereby seek to fulfil the requirements under the regulations;
- Apply consistent policies, procedures and controls across their group structure;
- Adopt appropriate internal controls within the business;
- Provide training to staff in relation to AML and CFT;
- Comply with new customer due diligence, enhanced due diligence and simplified due diligence requirements;
- Comply with requirements relating to politically exposed persons; and
- Ensure that record-keeping and data protection systems, policies and procedures are compatible with the requirements of the regulations and appoint anti-money laundering and compliance officers.

These requirements, together with those demanded by the various pieces of primary legislation, are extensive and demanding of firms' time and money,[21] but if implemented properly and fully they should theoretically enhance the capabilities of the regulated sector to both detect and prevent money laundering and terrorist financing. Failure to comply may result in a range of penalties, from warning letters to criminal prosecutions—a threat that is far from theoretical.

Suspicious Activity Reports

One of the themes of this study is how business has been mobilised by government to combat money laundering and terrorist financing. Business is assigned various roles within AML and CFT strategies, but among their most important tasks is providing information to national authorities.

If the business is regulated by the Money Laundering Regulations and there is a suspicion or knowledge of terrorist financing or criminal money laundering, the responsible person within the business must submit a suspicious activity report to the National Crime Agency (NCA) outlining the transaction and the cause or causes of suspicion. Specifically, anyone working in the business who comes across a suspicious transaction must report this to a nominated officer—a firm's Money Laundering Reporting Officer (MLRO). The MLRO will decide whether to send a suspicious activity report (SAR) to the UK Financial Intelligence Unit (FIU), which is located within the NCA but is operationally independent. The requirement for the regulated institutions to submit a SAR in relation to information that comes to them during the course of their business, if they know or suspect or have reasonable grounds for knowing or suspecting that a person is engaged in or attempting money laundering or terrorist financing, is to be found under POCA 2002 and TA 2000.

The NCA provides detailed guidance as to how to complete and submit a SAR, all of which is available on the NCA website. Submitting a SAR can also provide the reporter with a statutory defence to charges of money laundering or terrorist financing in relation to a future transaction of which they are suspicious.

It is thus possible for a business to make a Defence against Money Laundering (DAML) request in order to avail itself of the defence and ultimately proceed with the transaction. If a DAML request is refused within the 'notice period' (which is seven working days), law enforcement then has an additional 31 calendar days (the 'moratorium period')—from the day of refusal—to take further action, for example by restraining or seizing the relevant funds. If, however, in the case of criminal money laundering, after 31 days, the NCA has not re-engaged, the business may decide to proceed with the transaction. Changes adopted in 2017 have enabled law enforcement agencies to apply to extend the moratorium by an additional thirty-one days on up to six occasions (for a total of 217 days from the initial refusal by the NCA). Businesses need to be aware that the moratorium period is not applicable to terrorist-financing circumstances. In this case, the business will not have a defence until the NCA explicitly states that the transaction may proceed.

The overall number of SARs received by the FIU is staggering. Over the 18-month period from October 2015 to March 2017 (inclusive), the number of SARs submitted was 634,113. The 43,290 received in March 2017 were the most SARs ever recorded for one month.[22] In the SARS 2017 Annual Report it was noted that the banking sector remained the largest submitter of SARs, being responsible for 82.85 per cent of the total submitted.[23] There is now an existing database of over 2.3 million SARs.[24]

Given the significant consequences of becoming involved, albeit inadvertently, in money laundering or terrorist financing, or failing to properly report suspicions, many firms submit SARs defensively. That is, they will submit a SAR even in the most obscure of situations where the chances of money laundering or terrorist financing are limited. The feeling is that it is better to err on the side of caution than to miss a threat and become subject to harsh sanctions. Whether the FIU can properly assess all these disclosures is unclear, though it seems that SARs related to terrorist financing are given priority (see below).

The Criminal Finances Act 2017 has sought to reform the SAR regime by permitting financial and other credit institutions within the regulated sector to share SAR-related material with each other and then to submit to the FIU a joint SAR, known as a super-SAR. Those that avail themselves of this process will acquire defences against breaches of data protection laws and breaches of confidentiality requirements. The super-SAR process will fulfil both entities' disclosure obligations and will avoid the requirement for each entity to submit a separate report, but the process itself can be more complicated and longer. How the consequences of more extended moratorium periods, during which firms cannot act for their clients, will play out in practice remains to be seen. To these problems must be added the fact that the quality of many of the SARs submitted are reported to be variable thereby bringing into question the value of at least some of this product. It is fair to say that the SARs regime remains in need of reform.

Table 5.1: SARs Submitted by All Sectors, October 2015 to March 2017

October 2015 to March 2017	Volumes	% of total
Credit institution—banks	525,361	82.85%
Credit institution—building societies	22,323	3.52%
Credit institution—others	19,326	3.05%
Financial institution—MSBs	16,704	2.63%
Financial institution—others	23,675	3.73%
Accountants and tax advisers	6,693	1.06%
Independent legal professionals	4,878	0.77%
Trust or company service providers	112	0.02%
Estate agents	766	0.12%
High value dealers	265	0.04%
Gaming (including casinos) and leisure (including some not under Money Laundering Regulations)	2,223	0.35%
Not under Money Laundering Regulations (MLRs)	11,787	1.86%
Total	634,113	100%

Source: NCA, Suspicious Activity Reports (SARs) in the NCA Annual Report 2017.

Key United Kingdom AML and CFT Intelligence and Enforcement Organisations

There are two systems that are of relevance in the United Kingdom and reflect a dichotomy apparent in many jurisdictions that have developed relatively sophisticated AML and CFT systems. First, there is the enforcement system that implements the laws related to AML/CFT; and, secondly, there is the supervisory network that oversees and regulates selected parts of the national economy including in relation to AML/CFT.

National co-ordination and oversight of AML/CFT policies takes place at the level of the National Security Council. Below that the ministerial-level Criminal Finances Board (which brings together all relevant agencies) has responsibility for National AML policy development and implementation. At the policy level the Money Laundering Working Group (consisting of central government and law enforcement agencies) and the Money Laundering Advisory Committee (consisting of central government agencies, law enforcement agencies, regulatory supervisors and private sector organisations) oversee the United Kingdom's response to both money laundering and proliferation financing. The Terrorist Finance Board deals with terrorist financing issues. At the operational level the NCA led Criminal Finance Threat Group and the Economic Crime Threat Group meet quarterly and provide intelligence and operational co-ordination. Below this level various sub-groups meet on particular risk areas.[25]

Those organisations tasked with investigating and prosecuting money laundering and terrorist financing include the NCA, the police services of England, Wales, Scotland and Northern Ireland, the Regional Organised Crime Units (together with the Regional Asset Recovery Teams and the Asset Confiscation Enforcement Teams), the Serious Fraud Office (SFO), the HMRC Fraud Investigation Service, the FCA, the Crown Prosecution Service, the Public Prosecution Service of Northern Ireland (supported by the Economic Crime Unit), and the Crown Office and Procurator Fiscal Service (supported by the Economic Crime and Financial Investigation Unit and the Civil Recovery Unit). These organisations are in turn supported in terms of intelligence and analysis by the FIU, the NCA National Intelligence Hub (which identifies priority targets and produces relevant intelligence), HMRC's Risk Intelligence Service (which focuses on tax fraud), the SFO's Intelligence Unit (which focuses on complex fraud, bribery and corruption and related money laundering), the FCA, the police, the Joint Money Laundering Intelligence Task Force (JMLIT) (see below), the Security Service (MI5) and the Joint Financial Analysis Centre (a cross agency task force led by the NCA and HMRC originally dealing with the Panama Papers leak but now focusing more generally on money laundering methodologies and risks).[26] The Foreign and Commonwealth Office and the Office of Financial Sanctions Implementation (which is part of HM Treasury) manage financial sanctions and asset freezing.

The central organisation involved in combating serious and organised crime, including but not limited to money laundering, is the NCA, a non-ministerial government department. The NCA was created in 2013, the same year that the government published its Serious Organised Crime Strategy.[27] In 2017 the NCA identified high-end money laundering as one of the top-six national priorities for government agencies dedicated to combating serious and organised crime.[28] The NCA has the power under POCA 2002 to seize and forfeit cash, tax and undertake civil recovery of assets. It can demand organisations—which may be very unwilling to do so for client confidentiality and business secrecy reasons—to provide it with relevant documents and information.[29]

In its 2018 National Strategic Assessment, the NCA identified money laundering, fraud and other economic crime, bribery, corruption and sanctions evasion and cybercrime as amongst the key threats to the United Kingdom.[30] Within the NCA the National Economic Crime Centre (NECC) launched in October 2018 co-ordinates the country's response to economic crime and aims to reduce its impact on the United Kingdom's economy and society. It leads a multi-agency government effort to combat money laundering, and it also works with bankers and professional bodies to prevent criminal access to professional skills which could abet money laundering. Key departments and agencies located within the NCA and below the NECC include the International Corruption Unit, the International anti-Corruption Co-ordination Centre, the Civil Recovery and Tax Department, the Proceeds of Crime Centre, the JMLIT and the FIU.

Significantly, as noted, the NECC works towards building up the domestic coalition between government and the private sector, which is key to the proper functioning of the national AML and CFT effort, by co-operating with banks, professional bodies and regulators. To this end, in 2016 the JMLIT was established. Together with the British Bankers' Association and over forty major global and British banks, the JMLIT pools intelligence and expertise and targets high-end money laundering. According to the NCA, the joint taskforce has operational priorities which include understanding and disrupting trade-based money laundering; understanding and disrupting money laundering through capital markets; and understanding key terrorist-financing methodologies. The JMLIT also has a group dedicated to the identification of emerging threats.[31]

The importance of public–private sector coalition-building can be seen in the AML results claimed by the JMLIT. Thus between May 2016 and March 2017 JMLIT activities resulted in

- The arrest of 63 persons suspected of money laundering;
- The instigation of over 1,000 bank-led investigations into customers suspected of money laundering;
- The identification of more than 2,000 accounts not known to the authorities;
- The heightened monitoring by banks of more than 400 accounts;
- The closure of more than 450 bank accounts suspected of being employed for money laundering; and

- The restraint of £7m of suspected criminal funds.

We have already mentioned the FIU sits within the NCA but is operationally independent of it. The FIU attains its significance by being at the centre of SARs distribution analysis and dissemination.

We have already mentioned the FIU that sits within the NCA but is operationally independent of it. The FIU attains its significance by being at the centre of SARs distribution analysis and dissemination.

The FIU also works with other national FIUs including in the context of the Egmont Group in which global FIUs share information and co-ordinate in respect of AML and CFT. The NCA has given an example of how both the SARs system and international co-operation work. The example does not provide names of parties or countries, but it does state that a DAML SAR was received from a reporter describing a proposed transfer of funds between two countries. The reporter suspected that the account holder was wanted by overseas authorities for fraud and that monies in the account were in fact proceeds of fraud. The FIU refused the DAML request and held the funds during both the notice and moratorium periods. This activity permitted the foreign authorities to restrain the funds.[32]

A further example of the use of SARS in curtailing money laundering is where, following the emergence of the Panama Papers in 2016, the FIU undertook a daily analysis of SARs based on key word searches. The aim of this exercise was to reveal volumes, themes and trends in connection with the submission of SARs since April 2016 concerning Mossack Fonseca, the law firm at the centre of the scandal. The analysis resulted in weekly reports being sent to the NCA's Economic Crime Command senior leadership team which was able to assist foreign FIU's in relation to their investigations of Mossack Fonseca.

In the 2016–2017 NCA Report and Accounts later relayed in the *2017 National Risk Assessment*, the achievements of the NCA are elucidated. It is stated that NCA operational activities resulted in £82.8 million of funds 'being denied to criminals impacting on the UK' and the recovery of assets in the sum of £28.3 million. In addition, NCA activity also led to 1,441 arrests in the United Kingdom, and 1,176 arrests overseas 'across all crimes'.

Bear in mind, however, that the NCA's remit is broader than simply money laundering.[33] Turning specifically to counter terrorism financing there are three general objectives of UK CFT policy: (i) reducing terrorist-related fundraising in the United Kingdom; (ii) reducing terrorist finance moving in and out of the United Kingdom; and, most ambitiously, (iii) reducing terrorist financing overseas. In the United Kingdom, in support of these strategic objectives, the responsible department for CFT policy is the Home Office, which is supported by various other departments and agencies.[34]

Both the counter-terrorism and CFT structures are led by the Home Office and then overseen by the ministerial-level inter agency Terrorist Finance Board. This latter body sets out the CFT policy as expressed in the Counter-Terrorist Finance Strategy and the related Delivery Plan. The Counter-Terrorism Network enhances co-operation between various relevant agencies including in the CFT field.

In addition to the organisations outlined above in the context of general money laundering, government agencies involved in CFT operations are specifically supported by the National Terrorist Financial Investigation Unit (NTFIU) which is part of Special Operations 15 (SO15) of the Metropolitan Police (which takes the strategic lead for the country's counter-terrorism policing); Counter Terrorism Policing (an alliance of UK police forces working with intelligence agencies and supported by the 11 regional counter terrorism investigative and intelligence police units); and the Joint Terrorism Analysis Centre (which inter alia analyses intelligence related to terrorism, produces reports and sets threat levels).

Assessing AML and CFT Effectiveness

At the end of 2018 FATF delivered its long awaited Mutual Evaluation Report of the United Kingdom and provided the country with quite a glowing review and a generally clean bill of health. As John Glenn, the Economic Secretary to the Treasury noted, the 'report recognises that the UK's AML/CFT regime is the strongest of the over sixty countries assessed by FATF and its regional bodies to date'.[35] The report certainly noted that the United Kingdom had a robust understanding of money laundering risks and had improved significantly in the areas of national AML co-ordination and co-operation. Strengthened co-operation amongst law enforcement agencies, enhanced investigative tools, and mechanisms to enhance public-private information sharing were amongst the main very positive developments highlighted. The report stressed that 'JMLIT is an innovative model for public/private information sharing that has generated very positive results since its inception in 2015 and is considered an example of best practice' thus emphasising the importance of the public-private coalition.[36] The United Kingdom was also recognised as a global leader in promoting corporate transparency.

Particularly good results were also identified to have been achieved in the area of prosecutions. The report noted that the United Kingdom aggressively pursues money laundering investigations and prosecutions and that it had undertaken approximately 7,900 investigations, 2,000 prosecutions and 1,400 convictions annually for standalone money laundering or where money laundering was the principal offence. According to FATF, sentences appear 'to be effective, proportionate and dissuasive'.[37]

Good efforts were shown to have been made in preventing the misuse of legal structures and in cooperating domestically and internationally to address money laundering challenges. In terms of technical compliance the United Kingdom's legal framework was identified as being 'particularly strong' with only two areas requiring significant upgrades, one of which related to the FIU.

The identified weakness of the FIU was seen to stem from the fact that the authorities had made a conscious effort to restrict the role of the FIU in conducting operational and strategic analysis with the result that it was unclear whether SARS were being effectively exploited. Both the FIU and the SARS regime were identified as requiring an overhaul including the addition of human and technical resources. While some would question whether the FATF attack on the FIU was warranted or fair—certainly it would be possible to argue that the FIU remains one of the most powerful AML/CFT tools in the United Kingdom's armoury—it is clear what helped raise FATF inspector consternation. As the report notes:

> The volume of SARs received by the UKFIU is significant and continues to rise... However, the staff available to the UKFIU is inadequate with approximately 84 staff (currently around 80 full-time equivalent staff). At the time of the 2007 mutual evaluation, the UKFIU had 97 staff and expected an increase to 200 but it appears this surge did not occur.[38]

Promises made and promises not kept helped lead to this analysis. The report goes on to note that

> Only nine staff perform tactical analysis which is inadequate considering the increasing volume of SARs. The level of resources available for strategic analysis and international cooperation are also inadequate. The UKFIU uses a significant amount of its resources responding to DAML requests (27 staff), but there is a question about whether this is an efficient use of resources. Additionally, the UKFIU does not appear to have access to specialist skills such as forensic accountants to strengthen its ability to undertake sophisticated financial analysis.

While the FIU works more than competently in dealing with the challenges it faces, it is indisputable that more resources in this key area will enhance capabilities. The government responded quickly by noting that UKFIU staffing would increase in the short term by 30 per cent (with further increases planned in the future) and IT facilities would be improved.[39]

More problematically however is that the generally upbeat analysis has to be measured against the fact that the United Kingdom remains a centre for global money laundering activities. Against the criteria of FATF compliance the United Kingdom may be performing well but in the real world, challenges remain. Of course, one can always argue that without the efforts already undertaken the situation would be far worse and that is no doubt true. The FATF assessment makes a range of technical compliance suggestions which will no doubt enhance the UK's AML capability. But to radically upgrade the AML fight there will need to be political direction primarily aimed at high end money laundering and a willingness to invest significant resources into the relevant government AML agencies rather than just imposing more AML/CFT requirements on the private sector. Whether the political will is there for this fight is questionable.

The 2018 FATF report complemented the upbeat analysis of the United Kingdom's AML system with a very positive assessment of the CFT system. The report noted that the United Kingdom prosecutes a range of terrorist finance activity in line with its identified risks and that JMLIT has provided an important role in this activity. It notes that terrorist finance activity is properly integrated into broader counter terrorism strategies and that there is solid co-operation amongst the relevant agencies. It is emphasised that:

> Notably, counter-terrorism financing authorities have a close and fruitful relationship with both financial institutions and the non-profit organisation (NPO) sector. All [terrorist financing] convictions are subject to an expectation of imprisonment. The UK has demonstrated its ability and willingness to use all available measures to disrupt [terrorist financing], including freezing, seizure, and confiscation, as well as the removal of legitimate benefits and entitlements, orders to restrict activity and movement, and new powers which permit the seizure of funds in bank accounts.[40]

The NCA 2017 Suspicious Activity Report notes that during the 2016–17 reporting period, 2,026 SARs were identified and disseminated to the NTFIU and the Regional Counter-Terrorism Units network. This process involved the 'targeted review' of 26,655 SARs, identified from all SARs received. Given that there were 634,113 SARS submitted during the period surveyed, this seems to suggest that terrorist-related SARs made up 0.31 per cent of the total (or 4.2 per cent if one uses the SARs which were part of the 'targeted review').[41] Significantly, an officer of the NTFIU is seconded to the FIU in order to deal with terrorism related SARS in the most prompt and efficient manner.

Freezing terrorist assets is an important part of CFT strategy. The Office of Financial Sanctions Implementation (OFSI), is responsible for implementing domestic asset freezes under TAFA 2010, ACTSA 2001 and the Counter-Terrorism Act 2008. We shall look in greater detail at OFSI in the chapter dealing with sanctions.

The *National Risk Assessment of Money Laundering and Terrorist Financing 2015* Report noted that as of 31 December 2014, £117,000 had been frozen across eighty bank accounts under TAFA 2010 and various other regimes and regulations—a not overly large amount[42] but probably in line with the low level nature of the terrorist financing risk that the country faces. More details were forthcoming in quarterly reports submitted by the Treasury to Parliament on its operation of the asset-freezing regime as mandated by UN Security Council Resolutions, including Resolution 1373. At the end of 30 June 2017 under TAFA 2010 the minuscule sum of £9,000 had been frozen in a total of six accounts. At this time there were twenty designations under TAFA of which six were groups (the Basque group ETA, the Colombian ELN, Hezbollah Military Wing, the Syrian-based Popular Front for the Liberation of Palestine General Command (PFLP-GC), PFLP and Shining Path) and fourteen were individuals. Of the individuals all were overseas and none were in custody in the United Kingdom.[43]

Another £70,000 had been frozen under the United Nations' ISIL–al-Qaeda asset-freezing regime, for which the United Nations has responsibility for designations and for which the Treasury, through OFSI, is responsible for compliance under the ISIL (Da'esh) and Al-Qaida (Asset Freezing) Regulations 2011. Zero amounts had been frozen under EU Regulation 2580/2001, which implements Security Council Resolution 1373 against external terrorist threats to the European Union.

In figures presented in the 2018 UK Mutual Evaluation Report, FATF notes that between April 2012 and March 2017 police used ATCSA 2001 powers to make 79 cash seizures totalling £495, 797 and following the amendments to ATCSA 2001under the Criminal Finances Act 2017 authorities had by March 2018 used the new powers 13 times. According to the figures since 2001 TAFA 2010 orders have been employed for 158 individuals resulting in the freeze of £151,000. To these funds must be added the £70,000 frozen under UNSCR

1267 and the nearly £10,000 under UNSCR 1373 mentioned above.[44] As a result FATF rates the United Kingdom as having a high level of effectiveness in respect of terrorist finance preventative measures and sanctions.

At the time of writing no asset freezes had been made under the Counter Terrorism Act 2008.

OFSI itself does not actually freeze the asset. Rather, it is the responsibility of the institution (or person) which holds the assets to ensure that those assets are frozen. It is institutions in the private sector that are responsible for this activity and it is their onus if they hold relevant assets to report that matter to OFSI.

What about convictions? The first thing to recognise here is that not many persons have been convicted under terrorist-financing offences. In a factsheet related to the Criminal Finances Bill published in November 2016, it was stated that since 2001, 62 persons had been charged under terrorism legislation (TA 2000) with terrorist-fundraising offences.[45] In the *National Risk Assessment of Money Laundering and Terrorist Financing 2015* Report it was stated that there had been 17 convictions under sections 15–18 of TA 2000 in the period between September 2001 and June 2014. The report notes that 'this is not indicative of the total number of terrorist financing instances that have been disrupted'.[46] This was explained by the fact that in cases involving terrorist financing as well as other crimes such as murder, which have more severe penalties, the authorities may decide to pursue the charges with the more severe penalties rather than those related to the financing of terrorism. It notes—rather defensively: 'In addition, non-terrorism legislation can also be used to disrupt terrorist financing activity.'[47] It is worthy of note that the *National Risk Assessment of Money Laundering and Terrorist Financing 2017* Report did not present updated figures. Again however, these low levels should be seen in the context of the low level of terrorist finance threat that the United Kingdom faces.

Figures in the 2018 Mutual Evaluation report are generally consistent with this above analysis. Between 2012 and 2017 108 persons were charged with preparations of terrorist acts and 68 convicted. During this same period 25 persons were charged with terrorist financing offences of which 18 were convicted. While terrorist finance offences carry a punishment of 14 years' imprisonment, a fine, or both (preparation of terrorist acts are punishable by life imprisonment), there is as noted an assumption of imprisonment in terrorist finance cases. Convictions for terrorist finance offences are rarely appealed successfully. Individuals convicted of terrorist finance offences are usually also sentenced to counter-terrorism monitoring and notification orders which limit movement. As FATF notes 'These factors increase the effectiveness, proportionality and dissuasiveness of available sanctions'.[48] FATF also notes approvingly that where securing a terrorist finance conviction is not possible, the authorities effectively use other measures including pursuing other criminal charges, pursuing civil penalties and employing broader counter-terrorism powers. Specifically, authorities will attempt to disrupt terrorist financing through freezing or seizing terrorist funds or assets. As a result, FATF concludes that the United Kingdom has achieved a high level of effectiveness in terms of investigation and prosecution of terrorism offences.

Both government agencies and the private sector devote many resources to the CFT task. The United Kingdom nevertheless faces what amounts to a low-level terrorist financial threat expressing itself in very small sums used to mount lone wolf or small group attacks or to fund travel or to send to terrorist confederates overseas. The costs of CFT are as emphasised in this study significant and much of it is passed to the private sector. Because of the low probability-high impact nature of terrorist attacks a simple cost-benefit analysis of CFT may however not be particularly useful in this context. Following the money may not be the most efficient method of catching or deterring terrorists but it forms part of a panoply of actions which narrow the terrorist's room for manoeuvre. Private sector allegiance to this approach remains motivated by a mixture of patriotism and fear of financial and reputational risk should something go wrong on their watch.

The Supervision of Compliance in the Regulated Sector

Central to national AML/CFT policy is keeping the private sector aligned with government policy in respect of AML and CFT functions. Failures on the part of the private sector are met with stiff sanctions imposed by supervisory authorities. A clear example of this is the case of Deutsche Bank and its activities in Russia. Between April 2012 and October 2014, as we discussed in Chapter 1, customers in the Moscow branch of Deutsche Bank AG transferred more than $6 billion from Russia through Deutsche Bank in London to various bank accounts located in Cyprus, Estonia and Latvia. They did this by way of more than 2,400 so-called mirror trades, a means which, according to the UK financial services regulator, the FCA, was 'highly suggestive of financial crime'.[49]

The FCA, whose role includes regulating banks and overseeing their AML and CFT structures and policies, found major deficiencies in Deutsche Bank's

AML framework. These included inadequate customer due diligence, the use of flawed country and customer risk-rating methods, inadequate AML policies, inadequate procedures and inadequate information technology infrastructure together with a failure to provide proper oversight of mirror trades booked in the United Kingdom by traders located in overseas jurisdictions.[50] The bank was fined by the FCA a total of £163,076,224 for failing to provide an adequate AML framework. It was the largest financial penalty for AML control failings ever imposed by the FCA. Deutsche Bank, which reportedly proved exceptionally co-operative during the investigation and threw itself wholeheartedly into a new remediation programme, was able to secure a 30 per cent discount on the financial penalty, which would have been £229,076,224 without the reduction.[51]

The FCA is but one of 25 supervisory bodies including the HMRC, the Gambling Commission and various professional bodies whose role it is, amongst broader supervisory functions, to supervise the regulated private part of the AML and CFT coalition. Together, these bodies oversee a range of firms including financial institutions, designated non-financial institutions, law firms, accountancy firms and casinos. HM Treasury is responsible for appointing these AML and CFT supervisors and for the regulations which set out their roles and give them powers to effectively monitor their respective sectors. Each supervisor adopts a risk-based approach in respect of their sectors with the aim being to, inter alia, ensure proper implementation of AML and CFT without impacting on business effectiveness. To these ends, they conduct onsite and offsite inspections.

Specifically, the supervisory authorities ensure that members undertake proper customer due diligence, forward SARs to the NCA, and implement appropriate controls. All supervisory authorities attend a forum, also joined by HM Treasury, the Home Office and the NCA, where ideas and best practices are discussed. In 2017 a 'supervisor of supervisors', the Office for Professional Body AML Supervisors (OPBAS), was appointed to oversee the consistency and adequacy of AML and CFT supervisory arrangements. OPBAS is located within the FCA and will assist in co-ordination of relevant information between the professional body AML supervisors, statutory supervisors, and also the various law enforcement agencies. The organisation will aim at enhancing the consistency of AML supervision amongst professional bodies, but it will not be involved in supervising law firms or accountancy firms.[52]

Given its role supervising banks and other financial institutions, the FCA is the supervisory body that is most relevant to this study. Its current risk model views retail banking, wealth management, private sector banking, wholesale banking and capital markets as sectors presenting the greatest money laundering and terrorist financing risks; retail lending and e-money are seen as presenting medium risks; pension and retirement income, life insurance and retail investments are viewed as lower risks.

In their review of the large banks in the United Kingdom, the FCA found 'weaknesses in governance, and longstanding and significant underinvestment in resourcing ... This has often led to an ineffective risk-based approach with poor due diligence and monitoring standards, particularly, for higher risk businesses.'[53] Among smaller firms the problems derive from an understanding of their AML and CFT obligations and relate to their ensuring that their responses are proportionate to the business challenges they face.[54]

The problem of Sonali Bank (UK) (SBUK) demonstrates several of these challenges facing smaller banks and also the role of the MLRO. Sonali Bank (UK) Ltd is a Bangladeshi bank incorporated in the United Kingdom. Of its shares 51 per cent are held by the Bangladesh government and 49 per cent by Sonali Bank Ltd, Bangladesh. The bank was established to assist Britain's Bangladeshi community sending remittances from the country back to Bangladesh. It also sought to provide various banking services to the British Bangladeshi community as well as access to London's financial markets for Bangladeshi-based investors. By 2016 SBUK had received various warnings from the FCA, including some about inadequacies at all levels of its AML programme, its SAR reporting requirements, the oversight of its various branches, its governance structure and its obligations in relation to customer due diligence and especially politically exposed persons. The bank's former MLRO, who had been in place since 2011, was also censured. The FCA noted that despite constant warnings from the bank's internal auditors, he had failed to put in place proper AML monitoring arrangements, had reassured SBUK's senior management that controls were working when they were not, and had failed to request more resources or recruit more staff.

In October 2016 the FCA fined SBUK £3.25 million while the former MLRO was fined £17,900. The bank was also prevented from accepting new customer deposits for 168 days. Following a substantive review of its retail banking services, the bank took the decision in September 2017 to cease retail deposit banking services other than remittances and foreign exchange business. The effect of this de-risking decision would be to reduce at least some of the demanding customer due diligence requirements that it would have to undertake in the deposit-taking business. Of course a particular customer base was directly impacted by this decision some of whom may have been pushed into the informal and unregulated sector. As the FCA noted: 'We expect all regulated firms to promote a culture that supports sound AML controls and impresses on all staff the importance of complying with them.'[55]

Where there are failures by the regulated sector in implementing proper AML and CFT policies and procedures, supervisors have several sticks with which they can ensure coalition compliance. These include (i) the removal of that organisation's authorisation to offer its services to the public—essentially expulsion from the sector, the most severe sanction of all; (ii) temporary suspension of membership; (iii) the imposition of a financial penalty, as with Deutsche Bank; (iv) a reprimand in the form of a formal written warning; (v) a requirement of the organisation to implement a remediation programme—a potentially expensive undertaking—including a restriction on conducting a specific business or service; (vi) a plan of action as part of a general capacity-building programme rather than part of a formal disciplinary programme; and finally (vii) a warning cautioning the firm against specific conduct.[56]

Since 2012, the FCA has resolved 14 AML and CFT cases in respect of 10 firms resulting in penalties totalling £343,346,924 and also against four individuals resulting in fines totalling £92,700. According to FATF the 'Imposing [of] sanctions on individuals is a positive feature which has acted as a clear deterrent for firm's employees'.

The FCA also reported in relation to its supervised entities that 1 million suspicious cases were identified internally by banks of which 363,000 were escalated to the NCA. Of the 427 million retail and wholesale customers in the UK 1.6 million were identified as 'high-risk' including 120,000 PEPs (the latter constituting 0.02 per cent of all customer relationships). The FCA also reported that in 2017 banks in the United Kingdom rejected more than 1 million prospective customers and cut ties with 375,000 existing clients. Clearly this too raises de-risking concerns of the type we analysed in respect of the closing down of correspondent relationships.[57]

As the Treasury emphasises: 'Overall, effective supervision should leverage enforcement action to motivate regulated businesses to comply with AML/CFT obligations and mitigate risk.'[58] Yet, the *National Risk Assessment of Money Laundering and Terrorist Financing 2015* assessed that the UK supervisory regime was inconsistent, especially in applying a risk-based approach and also in providing a credible deterrent. It is certainly hoped that the appointment of OPBAS will iron out some of these difficulties. It should also however be noted that the head of the AML Supervisors' Forum claimed in

March 2018 that levels of non-compliance have averaged five per cent or less of the entities reviewed in the 2015–17 reporting period—so, by the supervisor's own measures at least, there is room for optimism.[59]

FATF in its 2018 Mutual Evaluation Report on the United Kingdom also commented that supervision was inconsistent across regulated entities and that the intensity of the supervision needed to be made better in line with the risks identified in the national risk assessment. It also articulated the view that while the FCA had a good understanding of money laundering and terrorist financing risks at the national and sector levels it had yet to demonstrate that it had an accurate picture of such risks at the firm-specific level.

Overall the 2018 FATF Mutual Evaluation Report only gives the United Kingdom a moderate rating in respect of supervision—though this of course is taking into account supervisors other than the FCA. In relation to the FCA the FATF report states

> The FCA notes that its recent supervisory work has shown some encouraging signs. For example, some major banks have put in place significant remediation programmes and other major firms are becoming more innovative in their approach to AML/CFT compliance. However, the FCA does not collect data on numbers of findings from its supervisory activities. Consequently, it was unable to provide any statistics demonstrating that its supervision has had a positive effect on firms' compliance (e.g. through a reduction in the number of inspections findings over time).[60]

The US Approach to AML and CFT

Twelve days after the 9/11 attacks, President George W. Bush issued Executive Order 13224 which, inter alia, prohibited transactions with persons who committed or supported terrorism and froze their assets. 'Today,' said Bush, 'we have launched a strike on the financial foundation of the global terror network ... Money is the lifeblood of terrorist operations. Today, we're asking the world to stop payment.'[61] The following month, in October 2001, Congress enacted the USA Patriot Act, which aimed to enhance the US government's ability to detect and prevent terrorist activities and to disrupt the links between terrorists, criminals and corruption. Title III of the Patriot Act amended the Bank Secrecy Act 1970 (BSA)[62] and demanded that US financial institutions implement programmes to prevent money laundering and terrorist financing. In this manner, the American government was to set the tone for public–private coalition-building, which was soon to be echoed in the European Union and elsewhere.

Like the United Kingdom and European Union, the United States follows FATF guidelines and UN resolutions and conventions by interpreting them within its particular legal and regulatory framework. United States federal bank regulators define money-laundering transactions as those intended to achieve any of the following: (i) to disguise the source of a transaction's funds; (ii) to disguise the disposition of the transaction's funds; (iii) to obfuscate the audit trail with the aim of giving the impression that the provenance of the funds is legitimate; and (iv) to evade income taxes.[63]

The Money Laundering Control Act 1986 (MLCA) makes money laundering both a federal criminal and also a civil offence. Like POCA, the US legislation defines what constitutes money laundering. This involves specific intent and knowledge, and covers the following: conducting a financial transaction knowing that the property involved is the proceeds of criminal activity; transporting or transmitting monetary instruments internationally; and conducting a financial transaction involving property that represents the proceeds of criminal activity. Attempting any of these activities is also covered by the Act, as is aiding or conspiracy. Predicate crimes include drug trafficking, bribery, smuggling and terrorist financing.

If an institution is a non-US entity—say, for example, a British or an Australian bank—the MLCA may have significant implications for that business. If such a non-American institution engages in a financial transaction which takes place in whole, or only partly, within the United States, the legislation will impact upon its activities, and it may be liable to be charged for money laundering. Also, if the British or Australian bank in the course of the transaction takes ownership of a property in which the United States has an ownership interest, then it too may fall within the ambit of the legislation. If the banks maintain bank accounts with a financial institution based in the United States, then American law may have bite. In this manner US anti-money laundering measures effectively reach beyond American borders.

While the MLCA defines money laundering, the BSA as amended (including by the Patriot Act), together with its implementing regulations, is the main United States law on money laundering in that it provides authority to the Secretary of the Treasury, delegated to the Director of FinCEN, to impose AML programme requirements on financial institutions (broadly defined to include banks, money service businesses, casinos and other entities). Specifically, BSA requires that financial institutions—the financial sector bears most of the burden of the measures established by BSA—set up programmes involving several important elements. These consist of (i) the adoption of relevant internal policies and procedures to ensure compliance with BSA requirements; (ii) an employee training programme; (iii) the designation of a specific compliance officer; (iv) the establishment of an independent audit function to test and monitor the programmes; and (v) the implementation of a customer identification programme grounded on a risk-based approach.

The customer identification programme aims at enabling institutions to be reasonably confident that they know the identities of their customers. In certain higher-risk circumstances institutions are required to undertake enhanced due diligence of clients and additional monitoring. This, too, in general terms is similar to the UK and EU approach.

The BSA further requires regulated financial institutions to both maintain records and file SARs, which 'have a high degree of usefulness ... in the conduct of intelligence or counterintelligence activities, including analysis to protect against international terrorism'. For example, institutions must keep records of cash purchases of negotiable instruments and file reports of any cash transactions that exceed $10,000 on a daily aggregate amount. The institution needs to put in place policies to detect and assess suspicious activities such as money laundering, tax evasion and terrorism, and then if necessary to submit a report.[64] Again, there are clear parallels with the UK and EU approach (and that of many other jurisdictions). The BSA, however, also places a reporting requirement for US persons who have signature authority over, or a financial interest in, a foreign financial account with a value of more than $10,000 and demands that a report on such an account be filed with FinCEN once a year. Institutions such as money service businesses must register with FinCEN while other institutions must maintain customer identification programmes and file currency transaction reports. It is very important for US institutions to report on suspicious activity that could signal money laundering, tax evasion or a range of other criminal activities. The importance of the BSA compliance programme is signified by the fact that it needs to be set down in writing, approved by the board of directors and reviewed annually.[65]

The third piece of American legislation we will look at is the one that is most famous: the Uniting and Strengthening America by Providing Appropriate Tools Required to Intercept and Obstruct Terrorism Act of 2001, otherwise known as the Patriot Act. Draft legislation had been prepared during the Clinton administration for combating drug cartels, and it was in 2001 adopted for counter-terrorist ends. Passed in the immediate aftermath of the 9/11 attacks, it is a very broad Act that amends the BSA specifically with regard to terrorist financing and covers border security, surveillance laws aimed at terrorism, and financial regulations. It reinforces public–private coalition-building as well as American extraterritorial reach. It requires sharing of information between the government and financial institutions, a verification programme of customer identity, enhanced due diligence programmes in various circumstances, and the creation and maintenance of AML programmes within the financial services industry.

Title III of the Patriot Act touches directly upon foreign financial institutions and extraterritoriality. Section 311 authorises the Treasury Secretary to assess whether a foreign country, financial institution, bank account or set of transactions is a 'primary money laundering concern' and then to impose 'special measures'. Special measures involve (i) keeping records and filing reports on specific transactions; (ii) obtaining information on beneficial ownership of accounts opened in the United States; (iii) securing information about customers whose transactions are routed through a foreign bank's 'payable-through' account; (iv) similar information in respect of customers using a foreign bank's correspondent account; and (v) imposing conditions upon the opening or maintenance in the United States of a correspondent or payable-though account.

One key focus here is on correspondent banking, which is regarded as a particular vulnerability. There is also a prohibition (section 313) on establishing a correspondent account with a foreign shell bank as these are recognised as being particularly vulnerable to criminal and terrorist exploitation. Correspondent accounts are therefore not allowed with foreign banks that do not have a physical presence within any jurisdiction. Similarly, indirect relationships are also forbidden because of the problem of 'nesting', whereby an at-risk institution can access the American financial system by way of a correspondent relationship with a non-risk correspondent which in turn has a correspondent relationship with an American institution. It is fair to say that the hard line taken by American authorities in particular against correspondent banking was critical in the de-risking policies adopted by many banks in this area.

Significantly, of course, failure to comply properly with the Patriot Act may result in various actions against American banks, including regulatory enforcement, civil financial penalties and even criminal actions. Again, as in the United Kingdom and the European Union, the firepower of the state is first turned against its own financial sector so as to ensure co-operation and bolster coalition efforts.

Section 314 of the Patriot Act further sets out the requirements for coalition-building between private financial institutions and state law enforcement as well as regulators, with a premium placed upon sharing information in regard to criminal money laundering and terrorist financing. Section 319(a) permits forfeiture from US correspondent accounts, while under section 319(b) law enforcement agencies are given enforcement powers by way of summons and subpoena authority in relation to foreign banks that maintain correspondent accounts in the United States. It is in this context that HSBC, with headquarters outside the United States, was fined nearly $2 billion in asset forfeitures and civil money penalties for permitting drug proceeds to enter the American financial system, while BNP Paribas was fined nearly $9 billion and suspended for a year for dollar-clearing services through its New York branch and other affiliates.

So much for the statutory architecture.[66] What about the institutions that arise from these laws and serve as the AML and CFT detection, prevention and enforcement mechanisms? The key point to recognise here is that in the United States 'the institutional framework for AML and CFT is complex, multi-faceted and involves a significant number of authorities from a range of ministries'.[67] As with the EU and UK approaches, there are both law enforcement and regulatory structures.

First, there are the main policy-making bodies, which are the core departments of state. The lead agency is the Treasury, which is responsible for issuing BSA regulations, which are then administered by the Treasury Department's Financial Crimes Enforcement Network (FinCEN). Federal bank regulatory agencies issue co-ordinating regulations which require banks to implement AML programmes.[68] The Department of Justice (DOJ) has prime responsibility for investigating and prosecuting relevant offences. DOJ is supported by the Department of Homeland Security (DHS), which too investigates money laundering and works to prevent terrorism. Other important departments include the State Department and the Department of Commerce whose roles we will further investigate when we deal in later chapters with export controls and proliferation of weapons of mass destruction.[69]

In addition, there are several other important policy-making bodies in the United States which have AML and CFT responsibilities. At the highest level the National Security Council deals among its many tasks with illegal financial threats related to terrorism, proliferation and organised crime. The Office of National Drug Control Policy develops the national drug control strategy and also has a money-laundering remit. The Office of Terrorism and Financial

Intelligence (TFI) is located within the Treasury, and develops and also implements AML and CFT strategies.[70]

At the level of operations, arguably the most important body is FinCEN, which acts as the United States' financial intelligence unit, sits within the Treasury's TFI and serves a regulatory function by administering the BSA in terms of ensuring that institutions are properly instituting AML and CFT programmes. It is empowered to take enforcement action against financial institutions for violations of these requirements (as are also federal regulatory agencies).[71] Clearly it has a much broader role than the United Kingdom's FIU.

FinCEN is the institution that receives and analyses SARs and plays a role in the Treasury's Office of Intelligence and Analysis, which is located within the Treasury's TFI and which is the Treasury's in-house intelligence unit. The TFI focuses on the financial infrastructure of terrorist groups, proliferation networks and organised crime groups. In turn, it is supported by the National Counterterrorism Center which integrates all sources of intelligence connected to external terrorism, including financial aspects. Also relevant in AML and CFT operations are the Department of Homeland Security Office of Intelligence and Analysis, the Special Operations Division within the DEA, the Department of State's Bureau of Counterterrorism, and the National Counterproliferation Center. These groups are in turn supported by various law enforcement agencies, including the FBI, the DEA, and the Bureau of Alcohol, Tobacco, Firearms and Explosives.[72]

The United States also has several regulatory organisations at the federal and state level with AML and CFT oversight of the financial sector. These include FinCEN, the Board of Governors of the Federal Reserve System and the Federal Banking Agencies.[73] The central federal authorities tasked with AML and CFT prosecutions include the US Attorney's Office, the Asset Forfeiture and Money Laundering Section of the Criminal Division, and the Counterterrorism Section of the National Security Division. This indeed is a large and complex AML and CFT architecture.

Measuring the Effectiveness of the US Approach

The 9/11 Commission noted that the existing financial regulatory AML framework did not fail in respect of 9/11 because it had been designed to detect anomalous transactions associated with global drug trafficking and financial fraud and not the routine type of transactions conducted by the terrorists.[74] Since that day, billions of dollars have been invested in upgrading the United States system for both AML and CFT purposes, and so the question arises as to how successful these massively expensive and complex approaches have been. As with the situation in the United Kingdom and European Union, it is difficult to measure such effectiveness with any clear-cut resolution.

One method is to gauge the views of those involved in the programmes. For example, in 2014 David S. Cohen, who as Undersecretary for Terrorism and Financial Intelligence in Treasury played an important role in strengthening the American AML and CFT system, noted that the combination of a powerful AML and CFT legal framework and an effective supervisory system has succeeded in making it more difficult for terrorists to access the US financial system, in many cases pushing terrorist support networks to resort to costlier and riskier means of meeting their operational needs.[75]

Other assessments to some extent support this optimistic assertion. The FATF 2016 Mutual Evaluation Report (the 2016 Report) on the United States suggests that the investment of money and effort is bearing fruit. It notes: 'The AML/CFT framework in the U.S. is well developed and robust',[76] and generally scores highly on FATF Immediate Outcomes. It remarks that co-ordination and co-operation both within the government and between public and private sectors are sophisticated and have matured significantly over the previous decade. It complements both the 2015 money laundering and terrorist-financing risk assessments. The 2016 Report goes on to note: 'The national AML/CFT strategies, key priorities and efforts of law enforcement and other agencies seem to be driven by these processes and are co-ordinated at the Federal level across a vast spectrum of agencies in several areas.'[77] It also emphasises that the American AML and CFT regime, in terms of effectiveness, achieves high results in the areas of prevention, investigation, prosecution and sanction for terrorist finance and proliferation finance and for preventing the exploitation of the non-profit sector. The 2016 Report points out that the US also achieves high scores for domestic co-ordination, international co-operation, the employment of financial intelligence, and investigating money-laundering offences 'such that only moderate improvements are needed in these areas'.[78]

A case in point is that of Saifullah Anjum Ranjha, who worked in an unlicensed money transmitter business that was involved in transmitting monies to overseas terrorist groups. A major sting operation was mounted by a range of co-operating law enforcement organisations including the FBI, US Immigration and Customs Enforcement and Homeland Security Investigations, the IRS Criminal Investigation Division, and the US Attorney's Office. Over $2 million was provided to Ranjha and his associates for them to transfer abroad to the Taliban, al-Shabaab and al-Qaeda. In 2008 Ranjha was convicted and pleaded guilty to conspiring to launder money, concealing terrorist financing and operating an unlicensed money transmission business. As a result, he was sentenced to over nine years in prison.[79]

This is just one example of a robust AML and CFT detection and prosecution system. The American authorities at the federal level secure approximately 1,200 money-laundering convictions a year under AML laws and regulations. As the 2016 Report notes: 'Many of these cases are large, complex, white collar crime cases, in line with the country's risk profile. Federal authorities have the lead role in all large and/or international investigations. There is however no uniform approach to State-level AML efforts and it is not clear that all States give [money laundering] due priority.'[80] The 2016 Report reveals that FinCEN receives approximately 55,000 new reports a day, including about 4,800 SARs. Between 2012 and 2014, the average number of SARs received each year was 1,725,322 while the average number of currency transaction reports was over 15,000,000.[81] The number of money-laundering charges and conviction rates under various laws stood at 3,369 and 1,967 respectively.[82]

In terms of CFT specifically, this unsurprisingly remains the major priority of the US authorities, as emphasised in the 2015 US National Security Strategy. The 2016 Report claims that whenever law enforcement agencies pursue a terrorist-related investigation, they conduct a parallel investigation to identify sources of financial support. The United States has succeeded in freezing significant amounts of terrorist funds 'and appears also to have kept terrorist funds out of its financial system to a large extent'.[83] Between 2010 and 2014 the authorities convicted approximately 130 individuals on one or more terrorist finance-related offences. Successful prosecutions included the Holy Land Foundation case, where funding was provided to Hamas, and various other fundraising cases involving al-Shabaab, al-Qaeda, Afghanistan and foreign terrorist fighters.

As with the situation in the United Kingdom and the European Union, the question still arises as to what these figures mean beyond their apparent size. How do you measure the effectiveness of these figures against the target of reducing money laundering and terrorist financing?

It is interesting to note that the 2016 Report still refers to 2010 UNODC figures estimating that the proceeds of financial crime in the United States (excluding tax evasion) was $300 billion or about 2 per cent of the American economy. By 2016 these figures were already six years old and no attempt was made to reflect any pattern or direction to support an argument of enforcement and regulatory efficacy. In relation to terrorist financing, while the 2016 Report states that terrorist funds have been kept out of the American financial system, how do we know this reflects the true picture? Could it be that, similar to the inability of the financial system in 2001 to identify the al-Qaeda attackers, the nature of the contemporaneous terrorist threat—which in many cases requires small amounts of funds and can make ready use of the informal sector—may again elude CFT mechanisms? It therefore remains challenging always to establish a clear link between all these efforts by both the public and private sectors and the prevalence (or lack thereof) of terrorist activity. Nevertheless, the American approach represents the gold standard for national responses and has extraterritorial impact well beyond US borders.

Responses to Corruption and Bribery

An analysis of national efforts to combat money laundering and terrorist financing needs to be accompanied by a complementary analysis of the specific steps undertaken by national authorities to counter bribery and corruption. In Chapter 1 we noted the links between bribery and corruption and money laundering; and the threat these activities pose for the integrity of global and national financial systems and the challenges they present for foreign trade and security policy. In Chapter 4 we reviewed global approaches to corruption and bribery in the form of the UN Convention against Corruption and FATF's efforts to link AML and CFT measures to these particular threats.

At the national level, two pieces of legislation have been key to combating bribery and corruption: in the United States there is the Foreign Corrupt Practices Act 1977 (FCPA) and in the United Kingdom the Bribery Act 2010 (Bribery Act).[84]

The FCPA was enacted in the late 1970s in order to help curtail the bribing of foreign officials, a practice that, while not endemic to American trade, was nevertheless taking place at a level that was clearly undesirable. At that time more than 400 corporations had admitted to using bribes to obtain business. Sectors of US industry involved in bribes reportedly included drugs and health care; oil and gas production and services; food products; aerospace, airlines and air services; and chemicals. For example, the Lockheed Corporation was involved in a major bribery scandal in Japan that tested relations between that country and the United States. In another example, Prince

Bernhard of the Netherlands was forced to resign from his official position as a result of an inquiry into allegations that he received $1 million in pay-offs from Lockheed. In Italy, alleged payments by Lockheed, Exxon, Mobil Oil and other corporations to officials of the Italian government eroded public support for that government and strained US relations with that country.[85]

The FCPA is jointly enforced by the Department of Justice and the Securities and Exchange Commission. There are two main provisions: those relating to anti-bribery and accounting. The anti-bribery provisions prohibit the payment of money or anything of value directly or indirectly with corrupt intent to a 'foreign official', political party, political party official, a candidate for political office, and foreign businesses and persons for the purpose of obtaining or retaining business. The accounting provisions relate to the requirements for keeping proper records of these transactions.

Violations of the FCPA can lead to significant fines. Breach of the anti-bribery provisions can result in individuals facing criminal penalties of up to $250,000 per violation and 5 years' imprisonment. The accounting and record-keeping provisions can result in individuals facing criminal penalties of up to $5 million for each violation and 20 years imprisonment. The imposition of other legislation may result in the fine being upto twice the benefit that the defendant sought to secure by way of the corrupt payment. The Securities and Exchange Commission may impose civil penalties of up to $10,000 for each violation and also require disgorgement.

Donald Trump, prior to becoming President, described the FCPA as a 'ridiculous' and 'horrible' law that made it difficult for American companies to compete overseas.[86] The FCPA nevertheless remains good law. In any event, by some measures at least, it is less expansive than the UK Bribery Act.

The Bribery Act was introduced into British law so as to address the requirements of the 1997 OECD Anti-Bribery Convention and to reduce corruption and bribery in international business transactions. At the time it was introduced UK policy in this space was regarded as out of date and not fit for purpose. Significantly, the Bribery Act has impact upon both companies incorporated in the United Kingdom which are operating abroad and overseas companies with a presence in the country. So US firms with a UK presence will need to take into account the domestic legislation in addition to the FCPA.

The Bribery Act covers active bribery and, notably, passive bribery (that is, receiving a bribe). There is a specific offence covering the bribing of a foreign public official with the intention of influencing that person in such a capacity.

Finally there is a corporate offence of failure by a business organisation to prevent active bribery. There is a strict liability on companies to prevent active bribery, the key defence being the existence of adequate procedures aimed at preventing persons associated with the company from engaging in bribery.[87]

Thus a UK incorporated company may be liable if one of its employees, agents or subsidiaries (so-called associated persons) bribes someone anywhere in the world for the purpose of securing or retaining business or related advantages. Concomitantly, a foreign company which is active in the United Kingdom will be liable for bribery even if this bribery takes place outside the country.

There are thus many similarities between the US and UK anti-corruption Acts, including their extra-territoriality, but there are also important differences, and the private sector has to ensure compliance with both countries' legislation. As noted, the demands of the British Act, in several places, exceed those of the older American law.

One of the key differences is that the Bribery Act covers both the giving and receiving of bribes, while the FCPA only covers the payment of bribes. Nor does the FCPA cover private bribery, focusing as it does on politically exposed persons (PEPs,) (though private bribery may be covered in other US legislative acts) while the UK legislation is also concerned with business-to-business commercial bribes. The FCPA requires proof that the person offering a bribe must do so with 'corrupt' intent while the Bribery Act has no such requirement in respect of the bribery of a foreign public official (though such a requirement is to be found in the general bribery offence). The FCPA has an exemption for small facilitation payments (sometimes called 'grease money') which are paid to officials to facilitate official actions. There is no such exemption under the Bribery Act although the British authorities have noted that prosecutors will exercise discretion in determining whether to prosecute in this regard. Furthermore, the FCPA permits promotional expenses to the extent that they represent reasonable and bona fide expenses. This is not the case under the Bribery Act.[88]

Firms seeking to navigate these countries' requirements will thus need to take into account several factors. These include the necessity to: review promotional expenses, and gift and hospitality guidelines; set in place strategies to reduce facilitation payments; expand their anti-corruption programmes so that they cover persons 'associated' with the company; and ensure that adequate procedures are in place to prevent corruption—a complete defence under the Bribery Act and an important protection or means of sentence mitigation elsewhere.[89]

It is interesting to note that in 2018 the FCA reported that it had discovered problems in organisations' anti-bribery and corruption controls, possibly due to the fact that firms have tended to focus their efforts on AML. The United Kingdom nevertheless continues to step up measures against PEPs and other criminals. In the Criminal Finances Act 2017 authorities are now permitted in various circumstances to apply for a so-called unexplained wealth order in relation to PEPs or someone connected with a serious crime who may own property inconsistent with their known sources of income.

The FCPA and the Bribery Act aim to a large degree at isolating overseas business from the threat of corruption, though such overseas activities also threaten unintended consequences in the home economies through the laundering process. Another mechanism by which the United Kingdom in particular seeks to prevent overseas corruption from spilling into local markets is by way of dealing with the requirement for increasing transparency in overseas jurisdictions.

Since 2016 both UK companies and limited liability partnerships must declare, in a searchable public register, the names of any individuals who retain a controlling interest of more than 25 per cent.[90] Such a requirement has traditionally not extended to Britain's overseas jurisdictions. We noted in Chapter 1 the linkages on the one hand of money-laundering integration into the London property market and on the other hand the connection of laundering processes to less-than-transparent company ownership structures emanating from several British overseas territories. Indeed, more than half of the companies mentioned in the Panama Papers were set up in the British Virgin Islands.

Pressures for establishing such public registers where beneficial ownership of company structures can be readily determined have met with opposition from several of these territories. In early 2018 the British Virgin Islands (BVI) government informed British MPs considering such a move that a public register 'effectively disenfranchises the BVI legislature and its electorate creating resentment and damaging the historically strong relationship between the BVI and the UK'.[91] The fear was that calls for public transparency would deter investors seeking anonymity (the vast majority of whom do so for legal and not money-laundering reasons), cost tens of millions of pounds, and endanger overseas territory economies. They could also point to the fact that these territories had already—albeit under some pressure—introduced private central registers of ownership which, while not totally public and transparent, were accessible to British tax authorities and law enforcement.

These arguments ultimately held little sway, and in the Sanctions and Anti-Money Laundering Act 2018 the British government stated that for the purposes of detecting, investigating and preventing money laundering, it will assist British Overseas Territories in establishing publicly accessible registers of beneficial ownership of companies registered in those jurisdictions. The territories most likely to be affected by these measures are the British Virgin Islands, Cayman Islands, Bermuda, Anguilla, Turks and Caicos Islands, and Montserrat.[92]

Forcing companies to reveal their beneficial ownership will undoubtedly play a role in stemming the inflow of laundered funds into the London property market. That overheated market will probably benefit from this move, though it remains to be seen how it will affect overseas territories that have based significant elements of their economies on attracting investors who for various reasons require secrecy about company ownership structures.

Other efforts aimed at enhancing transparency including the aforementioned register of controlling interests, a beneficial ownership register of both United Kingdom and foreign trusts with tax consequences in the United Kingdom (not publically open but available to law enforcement agencies) and the planned public register of beneficial owners of non-United Kingdom entities that own or purchase property in the United Kingdom or participate in United Kingdom government procurement will, it is hoped, support the fight against bribery and corruption in the United Kingdom.

6

FINANCIAL SANCTIONS
AND EXPORT CONTROL REGIMES

Anyone doing business with Iran will NOT be doing business with the United States. I am asking for WORLD PEACE nothing less.

President Donald Trump, Tweet, 7 August 2018

Iran versus Iraq

In the second year of his war against Iran as the Iranian army began to develop its 1983 Val Fajr II campaign near Haj Umran, Saddam Hussein ordered Iraqi forces to begin firing mustard gas ordnance against enemy mountain-top positions. Buoyed by the potential of these chemical attacks and in response to the beginning of major Iranian counter-offensive operations in the Majnoon Islands not far from Basra, in early 1984 Iraqi forces started augmenting its mustard gas strikes on Iranian infantry groups with employment of the nerve agent Tabun.[1]

It was shortly after these initial Iraqi chemical assaults that Mohsen Rafighdoost, minister of the Iranian Islamic Revolutionary Guard Corps, met with the supreme leader, Ayatollah Khomeini, and briefed him on plans for Iran to begin work on its own chemical, biological and nuclear weapons. Khomeini's response was unequivocal: weapons of mass destruction were forbidden by Islam and no such programmes could be initiated. Not to be deterred, in June 1987, following an Iraqi air attack using chemical weapons against civilians in the Kurdish city of Sardasht, Rafighdoost once more approached the supreme leader requesting approval for chemical and nuclear weapons. Again he was rebuffed with the unequivocal retort: 'We don't want to produce nuclear weapons.'[2]

By the end of the Iran–Iraq War, however, Iran was regularly using artillery and other means to deploy chemical weapons against Iraqi positions. And soon after the war, any Iranian inhibitions on developing nuclear weapons collapsed as well. While Iranians were later to deny they had an interest in nuclear weaponry, they were in the late 1980s not as cautious in their proclamations. In October 1988, Akbar Hashemi-Rafsanjani, then Speaker of the Iranian Parliament and Commander-in-Chief of the Iranian forces, addressed the Iranian military with the words:

> With regard to chemical, bacteriological and radiological weapons training, it was made very clear during the war that these weapons are very decisive. It was also made clear that the moral teachings of the world are not very effective when war reaches a serious stage ... We should fully equip ourselves both in the offensive and defensive use of chemical, bacteriological and radiological weapons.[3]

As late as 2003 and possibly later, Iran had a developing set of uranium and plutonium programmes dedicated to deploying a nuclear weapon.[4] In response to Teheran's proliferation efforts, rather than engaging in preventive military actions, which seemed militarily difficult and politically unappealing, Western countries led by the United States began to rely primarily on financial and trade sanctions as well as selected embargoes to control Iran's nuclear ambitions.

The imposition of financial and other sanctions on Iran, as well as US extraterritorial (secondary) sanctions targeting non-American financial institutions willing to transact with the Iranian regime, played a critical role in forcing Iran to enter into the Joint Comprehensive Plan of Action (JCPOA) in 2015, whereby, in exchange for phased sanctions relief, Iran promised to freeze various elements of its nuclear development programme. The reimposition of sanctions following Washington's withdrawal from the JCPOA in 2018 emerged as a critical component behind efforts of the Trump administration to force Teheran into a better nuclear disarmament deal and as a means for securing improved behaviour by a country identified by the United States as a major state supporter of terrorism. Again, secondary US sanctions targeting non-American businesses emerged as a vital element of this strategy.

This chapter will start with an analysis of the scope and key elements of the United States' financial and trade sanctions and then focus on the extraterritorial aspects of this sanctions regime. It will show how the goal of ensuring and policing a public–private partnership, necessary for the effective prosecution of sanctions, is, in the case of the United States, a policing function extending beyond its national borders. This analysis will be accompanied by a review of the different sanctions strategy adopted by the European Union. The Iranian case will demonstrate how tensions have emerged between these two power centres, especially in respect of efforts by the Europeans to resist American secondary sanctions targeting its financial institutions. It will argue that the Iranian case reflects both the convergence and the conflicts underpinning the different sanctions strategies and the difficulties this poses for financial institutions trying to navigate around sometimes conflicting national demands. The chapter will then move on to an analysis of US and EU military and other embargoes and the integration of those embargoes with financial and trade sanctions policies. Finally it will analyse the complementary roles that both sanctions and embargoes have played in response to Russia's invasion of eastern Ukraine and its annexation of the Crimean Peninsula.

Financial Weapons of Coercion and Coalition-building

As the example of Iran clearly demonstrates, sanctions are in their most aggressive expression financial weapons of coercion. They differ from the anti-money laundering (AML) and counter-terrorist financing (CFT) tools in that they do not primarily involve adopting defensive mechanisms to protect the integrity of a state's financial system but rather aim to take the fight to the enemy, forcing it to alter its behaviour by the imposition of financial, economic and other measures of hardship. While sanctions may take the form of travel, sporting and various bans, it is the restriction of financial services and trade relations that has the greatest bite and in many cases the greatest potential for forcing changes to the targeted policies. The scope of sanctions may vary considerably. Ultimately this depends upon the foreign and security goals of the party seeking to employ them, who the employing party is (the United Nations, a regional body such as the European Union, or a state such as the United States or United Kingdom), the type of target (states, companies or individuals) and the specific behaviour of the target that needs amending.

Since the end of World War II sanctions have been increasingly used, often as a substitute for military action and often with mixed results. Article 41 of the United Nations Charter refers to the power of the Security Council to authorise the employment of sanctions as a response to a threat to international peace and security. Specifically:

The Security Council may decide what measures not involving the use of armed force are to be employed to give effect to its decisions, and it may call upon the Members of the United Nations to apply such measures. These may include complete or partial interruption of economic relations and of rail, sea, air, postal, telegraphic, radio, and other means of communication, and the severance of diplomatic relations.[5]

UN sanctions are binding on all its members, with the most common sanctions being asset freezes, travel bans and arms embargoes. The organisation sets sanctions objectives which members should support and integrate into their local laws. Currently there are over a dozen UN sanctions regimes focusing on supporting counter-terrorism, non-proliferation and the political settlements of conflicts. Regimes are each administered by a sanctions committee supported by monitoring groups, panels and teams.[6] There are various sanctions lists and a Consolidated UN Security Council Sanctions List, which includes the names of all entities and individuals subject to sanctions imposed by the Security Council.[7]

The effectiveness of these sanctions is subject to the inherent political limitations of the organisation and the fact that the United Nations has no direct means of enforcement. Over the past few decades, however, it is the United States and the European Union that have taken the lead in sanctions policy.

The expansion of the use of sanctions, together with supporting export control regimes, as instruments of foreign and security policy has impacted on a range of industries within the sanctions-implementing states but especially on that of finance and banking. The co-option of the private sector into a public–private coalition as a key partner in support of sanctions and embargo policies has placed enormous strains on the financial sector, as institutions need to add to the already pressing demands of national AML and CFT policies the need to accommodate guidelines aimed at affecting the behaviour of targeted states and entities.

Global financial institutions are essentially ill suited for such a role. Furthermore, the nation states within which these financial institutions operate may have different and conflicting approaches to sanctions and targets in line with their unique foreign and security policies. These inconsistencies, together with the fact that the sanctions and embargo landscape is constantly changing, appear sometimes of little consequence to national leaderships, which impose huge penalties on financial institutions that can find themselves, often unwittingly, engaging in or supporting transactions that run counter to current sanctions policies. American penalties have been significant in this context, not only because of their size but because of their extraterritorial reach. European banks—all incorporated outside the United States—have been the primary target of Washington's efforts to keep the public–private coalition in line. Since 2012 three European banks have each been fined billions of dollars for various American sanctions and export control breaches. In 2014 BNP Paribas was fined $8.96 billion; in 2012 HSBC Bank was fined $2.29 billion; and in 2015 Commerzbank AG was fined $1.45 billion. These are significant sums even for large-scale global financial institutions.

Converging and Diverging United States and European Union Sanctions Regimes

American sanctions have come a long way since the early nineteenth century when the first target was Great Britain, which the United States alleged was harassing American sailors. Today, the United States boasts the most extensive and sophisticated global sanctions system. Sanctions originate in the executive and the legislative branches of the US government as well as at state and local levels. The organisation tasked with administering this sprawling sanctions regime against countries and groups of individuals is the Office of Foreign Assets Control (OFAC) which was created in 1950 following China's entry into the Korean War when President Truman sought to block North Korean and Chinese assets subject to American jurisdiction.[8] Companies may access a description of each sanctions programme or embargo in the Sanctions Program and Country Summaries area[9] and in the Regulations by Industry area[10] on OFAC's website. OFAC also grants licences to permit entities and persons to transact in circumstances which otherwise may be prohibited by a specific sanctions programme.[11] Types of licences relevant here include a general licence, which permits a particular type of transaction for a class of persons without the need to apply for a licence each time a transaction is undertaken; and a specific licence, which is a document issued by OFAC in response to an application, providing a specific entity or person authorisation to conduct a particular transaction.[12]

There are two key aspects of the US sanctions system that concern us in this study. First there is the size of the civil and criminal penalties. As OFAC points out, 'The fines for violations can be substantial. In many cases, civil and criminal penalties can exceed several million dollars.'[13] In fact, as we have already just noted, they can exceed billions of dollars and far outweigh any fines issued by European states enforcing their own sanctions. This reality has important implications for banks and other entities deciding upon which sanctions regime to give priority to—especially when those regimes conflict.[14]

The second most notable characteristic of the US sanctions regime is its long jurisdictional arm. There are three key subjects of US jurisdictional scope that are relevant here: (i) US persons ('US Persons'); (ii) non-US persons '(Non-US Persons') who are involved in what are for the Americans sanctionable activities or are engaged in helping cause violations of those sanctions; and (iii) anyone transacting in items of American origin or content. We will now deal with the first two of these subjects, with the third being covered in the following section on embargoes.

US sanctions prohibit US Persons from being engaged directly or indirectly with countries or persons subject to sanctions. For this purpose US Persons are:

- Entities organised under US laws and their non-US branches (that is, entities located outside the United States but incorporated within the country);
- Employees of these entities irrespective of their nationality;
- Entities and individuals physically located in the United States, even if such location is intended only to be temporary; and
- All US citizens and Green Card holders.

Concomitantly, Washington employs sanctions regimes of differing intensity. By the end of 2018 comprehensive US sanctions regimes targeted Cuba, Iran, Syria, Sudan, North Korea and Crimea; significant sanctions targeted, for example, Russia and Myanmar; and more limited sanctions regimes included those targeting Venezuela, Lebanon, Yemen and Somalia. There are also specific non-state entities that are targeted, which include drug traffickers, terrorists, weapons of mass destruction proliferators, and persons engaged in cybercrime. Such individuals and groups may be active within states or they may not be country specific. These individuals are referred to as specially designated nationals (SDNs). Sanctions prohibit US Persons from dealing directly or indirectly with SDNs and any entity 50 per cent or more owned or controlled by one or more SDNs. Nor can US Persons facilitate such transactions by non-US persons. Prohibited facilitation includes financing, negotiating and drafting relevant contracts, supplying services or products in support of business, supporting information technology, and amending procedures and policies to enable the transactions.[15] There is a further requirement to actively block the property of SDNs and any interests they may have in that property. Penalties can be extremely severe. They include not only unlimited fines but also criminal penalties and jail time, extradition, disqualification of directors of companies, revocation of licences and reputational damage.

Sanctions as a foreign policy and security tool are most effective when they have wide international support. If some states are to unilaterally restrain national financial and economic activity with a target while other states exploit resulting opportunities, this will undermine the relevant sanctions regime and cause domestic disquiet among investors and exporters in the jurisdiction where restraint is policy. There has to be both the perception and reality of a level business playing field.

Consequently, when we speak of extraterritoriality in the context of American sanctions, what we are referring to are measures that specifically target non-US entities and individuals and seek to control investment and trade activities by such entities outside American territory. The aim of these sanctions is to ensure a united international sanctions front and also to make certain that American institutions are not disadvantaged relative to foreign businesses in the pursuit of commercial opportunities.

What Non-US Persons (wherever they are located) need to be aware of is that certain of their activities may trigger American sanctions. These companies will be subject to sanctions if they engage in 'sanctionable activities' as defined by the United States. Thus they may become subject to some of the penalties we have already described if they, for example, deal with SDNs as well as entities 50 per cent or more owned by one or more SDNs. Non-US Persons will breach sanctions if they cause violations of American laws, for example by using dollars that enter the United States financial system (for example dollar clearing) to make payments to sanctioned targets, or by ordering American goods for the purpose of supplying SDNs.

The prohibition of the use of dollars, the world's major international currency, acts as a very important brake on Non-US Persons who would wish to trade with parties subject to American sanctions. The vast majority of dollar transactions will have a US nexus. So when, for example, an Indian technology company trading with a Teheran-based counterparty invoices that counterparty in dollars for the transfer of relevant parts, the financial element of the transaction could be routed through the American financial system. This is because the Iranian importer will order its Dubai bank (for the sake of argument, a bank which itself is not subject to US sanctions) to pay out of its Dubai account the Indian company but that Dubai bank will need to transact with a US bank with which it maintains a correspondent banking relationship and which possesses the necessary dollars. The US correspondent bank will then debit the relevant correspondent account and simultaneously credit an account belonging to another bank which maintains a correspondent relationship with the Indian exporter's bank in Delhi. The funds are then transferred to the Delhi account of the Indian exporter. It is this use of dollars and the American financial system that will lead the American authorities to claim jurisdiction. The focus on targeting external entities in allied states has raised considerable tensions with the European Union.

As part of its common foreign and security policy the European Union imposes 'restrictive measures' on various targets. Sanctions decisions take the form of regulations and must receive unanimous agreement from all EU members. They are then directly applicable in all member states. Sanctions are implemented for a defined period (usually a year or six months) following which they require renewal. As with the AML and CFT context, penalties and enforcement (and in this instance licensing) are implemented at the national level.[16] Since the early 1990s, the European Union has imposed sanctions more than thirty times, the most comprehensive of which have targeted Iran.

The European approach to sanctions has always been narrower and more targeted than that of the United States. The jurisdictional scope of EU sanctions cover: (i) entities incorporated in a member state and their EU and non-EU branches; (ii) all businesses operating within the European Union, including businesses by non-EU entities; (iii) individuals located within the European Union irrespective of their nationality; and (vi) individuals who are nationals of an EU state wherever they are located.

While Non-US Persons may violate US sanctions with the use of dollar payments, the European Union does not have a similar concept of violation of its sanctions if a party not subject to EU jurisdiction employs euro payments or otherwise generally breaches EU sanctions.

The European Union prefers to shy away from comprehensive sanctions of the type that Washington developed against Iran, preferring more focused or 'smart sanctions' which target, for example, specific sectors of an economy. So the European Union may target financial services, brokering and insurance and funds transfer; or may target various sectors of the economy such as oil and gas, defence and technology. Some of the smart sanctions may develop quite widely and essentially transform into comprehensive sanctions—for example, the approach to Syria, Crimea, North Korea, and Iran at least until the JCPOA in 2015. There is an extensive reliance on targeting designated persons in countries such as Ukraine, Afghanistan and the Balkans, as well as specific terrorist groups. There is also a strong reliance on arms embargoes, which we will deal with in the next section. As with American sanctions, the Europeans also seek to prevent circumvention and facilitation.

By the end of 2018, the European Union, which aims to align with UN multilateral measures and sometimes imposes its 'EU autonomous' sanctions, (for example in Ukraine and Syria) retained its own generally narrow restrictive measures on Egypt, Mali, Tunisia, Guinea and Guinea-Bissau, and joined with the United States in sanctions on Libya, Venezuela, Lebanon, Belarus, Ukraine, Crimea and Sevastopol, Syria, Haiti, Iraq, Russia, Iran, Afghanistan, China, North Korea, Burma, Eritrea, Yemen, Sudan, Serbia, Montenegro, Moldova, Bosnia and Herzegovina, South Sudan, Somalia, Burundi, Democratic Republic of Congo, Central African Republic and Zimbabwe. There are also EU restrictive measures targeting ISIS and al-Qaeda.[17]

While the United States employs the SDN category, the EU sanctions regime identifies Designated Persons (DPs). Specifically, EU members are required to freeze funds and economic resources owned or controlled by DPs, and they are prohibited from directly or indirectly making funds or tangible or intangible economic resources available to DPs, including payments. While in the United States the relevant assessment is whether the entity is 50 per cent or more owned by one or more SDN, the EU test is rather more complex and involves determining whether the entity in question is owned or controlled by a DP or DPs. Consistent assessments are not helped by the fact that each member state interprets such ownership and control according to its own laws.[18] How certainty can be reached in this area is not easy, and a risk-based approach underpinning a proper and recorded screening effort provides the best method for companies seeking to trade.

Within the United Kingdom, the sanctions regime consists of targeted asset freezes; restrictions on financial markets and services targeted against people, organisations or sectors; restrictions on access to capital markets; directions for ceasing banking relationships and activities; requirements to seek authorisation before payments are made or received; restrictions on the provision of financing, brokering, insurance or advisory services; and directions to cease all business. UK jurisdiction extends to anyone in the United Kingdom, and British citizens wherever located, any corporate entity operating in the United Kingdom, and any entity incorporated under the law of any part of the country, wherever located.

Responsibility for financial sanctions implementation falls on the Treasury acting through the Office of Financial Sanctions Implementation (OFSI).[19] Specifically, OFSI is responsible for the implementation and administration of international financial sanctions effective in the United Kingdom as well as licensing exemptions; and for the implementation and administration of domestic designation legislation.[20] In 2017–18 the United Kingdom implemented 29 financial sanctions regimes for reasons of combating terrorist financing, nuclear proliferation, human rights abuses and violations of national sovereignty.[21]

OFSI maintains two key lists. First, there is the Consolidated List of Financial Sanctions Targets in the United Kingdom (the UK Consolidated List), which includes all those persons subject to financial sanctions under EU and UK legislation as well as UN sanctions implemented through European Union regulations.[22] It does not include other European national-level sanctions or non-EU listings such as those of the United States.[23] On 3 April 2018, 2,077 persons and entities within 26 financial sanctions regimes were targeted and appear on the UK Consolidated List. As of September 2017, £12.8 billion of frozen funds were held by businesses in the United Kingdom (including funds of Libya, Iran, Syria, Egypt and Ukraine).[24] In addition there is also a Capital Markets Restrictions List, which contains a list of entities subject to specific capital market restrictions not found in the UK Consolidated List. The emphasis here is on Ukraine.[25] Note that the lists cover the entities with which a company should not be transacting (unless it has a licence or there is an exemption) and not what they are importing or exporting.

There are general and specific business and professional reporting obligations in respect of sanctions. In terms of the former obligation, everyone must provide OFSI with information of sanctions breaches and dealing with DPs; while in relation to the latter requirement, a range of groups have these reporting obligations including legal professionals, accountants, tax advisors, estate agents, auditors, casinos, dealers in precious stones and metals and trust or company service providers.[26] In the United Kingdom a breach of financial sanctions is punishable by fines and/or a prison sentence.[27] Monetary penalties may be the greater of £1,000,000 or 50 per cent of the value of the breach while sanctions breaches could result in imprisonment for up to 7 years. Countries within the European Union may each adopt different guidelines and punishments.

At the very least, businesses need to be aware of the content of the OFSI lists as they seek to determine with whom they may or may not transact and also for purposes of their reporting obligations. These lists, together with the US lists in particular, are necessary though not sufficient in helping answer questions such as whom companies may directly or indirectly receive funds from or pay funds to; what shareholding threshold is being applied to determine whether the contracting party is a DP or SDN; whether there is knowledge of the ownership structure of the contracting party; what level of due diligence can ever be sufficient here; and whether US comprehensive-sanctioned territories are in any way involved. OFSI encourages companies to check against the UK Consolidated List and 'if you're unsure and there is no physical risk to yourself, consider asking the individual or organisation for more information.'[28] OFSI emphasises that the company is responsible for exercising proper due diligence. In 2017–18 OFSI received 122 reports dealing with suspected financial sanctions breaches totalling approximately £1.35 billion.[29]

OFSI also acts to issue licenses which permit businesses and individuals to conduct transactions which would otherwise be prohibited under the financial sanctions regimes. Licensing grounds vary dependent upon the sanctions regime in question and may include requirements of paying legal fees, fulfilling prior contracts, paying extraordinary expenses or acting on humanitarian grounds. In 2017–18 OFSI issued more than 50 licenses the majority of which went towards payment of legal fees.[30]

OFSI does highlight some potential red flags that UK institutions should be aware of. These include: whether an individual or company matches (or partly matches) a name on the UK Consolidated List; when a counterparty employs overly complex payment structures; where a refusal by the exporter to do business with Company A is followed by a request from Company B (with a similar profile to Company A) for the same items or services; and where the entity receives requests for the same quantities of equipment at the same time from different companies or individuals.

These are, unfortunately, not always easy issues to resolve as the lists are extensive and the names included in the lists are sometimes unclear and appear malleable. How would a company determine with a great degree of confidence whether a contracting party is directly or indirectly controlled by a DP or SDN? How is one to definitively determine the beneficial and ownership structures of businesses given that there are often no objective searchable registers of interests and that structures are often far from transparent or simple?[31]

Iranian Sanctions and US–EU Tensions

Prior to the JCPOA implementation in January 2016, the United States had already been sanctioning Iran for decades for a range of reasons, stretching from Iran's support of terrorism to its development of nuclear weapons. The breadth of American laws dedicated to punishing Iran and changing its behaviour and the means employed (including asset freezes, and trade and investment bans) were extensive. Initially, the goal of American policy was broader than counter-proliferation, but as Iranian proliferation activities grew, Washington's policy began to focus primarily on Iran's nuclear programme.

The first US sanctions on Iran involved the freezing of Iranian financial and other assets within the United States and arose in response to Iranian students in 1979 storming the American embassy and taking Americans hostage. Iran also soon became subject to sanctions aimed at curtailing its broader terrorist-supporting behaviour. In 1983 Iranian-backed terrorist groups that ultimately formed into Hezbollah killed over 240 US Marines in their barracks in Beirut and led to Iran being designated by Washington as a state sponsor of terrorism. This 1984 designation resulted in various sanctions being imposed on Iran, including restrictions on sales of dual-use items, a ban on direct financial assistance and arms sales to Iran, a requirement that the United States oppose multilateral lending to that country, and the withholding of American foreign assistance to states and organisations that supported Iran.[32]

A focus on Iran's terrorist-supporting activities was ultimately accompanied by a focus on its emerging weapons of mass destruction programme as concerns about both Baghdad's and Teheran's nuclear ambitions became more acute. In 1992 Congress passed the Iran-Iraq Arms Non-Proliferation Act. Pursuant to that Act the United States forbade transfers of technology and goods to either Iraq or Iran in circumstances where there was a concern that such a transfer would support either state's acquisition of chemical, biological, nuclear or advanced conventional weapons.

From 1995 the Clinton administration broadened sanctions with several executive orders, including Executive Order 12959, which barred American trade with and investment in Iran. During the following year, Congress passed the Iran–Libya Sanctions Act (ILSA), which later became known as the Iran Sanctions Act. The Act penalised entities for investing in Iran's petroleum industry, as such investments could provide the country with funds to develop weapons of mass destruction and also to support terrorism. There was also an extraterritorial dimension to the Act, which permitted the United States to impose secondary sanctions on foreign companies that invested in excess of $20 million a year in Iran's oil or gas sector.

Following 9/11, various executive orders gave detailed effect to US sanctions measures. Executive Order 13224 (2001) authorised the government to block assets of designated foreign individuals and organisations involved in terrorism and the support of terrorism. Later orders specifically targeted Iranian activities in the fields of both terrorism and proliferation. So certain financial institutions such as Bank Melli and Bank Saderat, their subsidiaries and branches, were designated and US Persons were forbidden to deal with them. For example, General Qassem Suleimani, head of the Iranian Revolutionary Guard Corps—al-Quds Force, was also designated for a range of undesirable activities. Not only were US Persons required to steer well away from these organisations and individuals, but so too were international businesses. As a 2007 Treasury press release stated in respect of various institutional and individual designations issued under Executive Order 13382 (2005): 'Today's designations also notify the international private sector of the dangers of doing business with three of Iran's largest banks, as well as the many IRGC-affiliated companies that pervade several basic Iranian industries.'[33]

While the United States never lost sight of Iran's terrorist-supporting behaviour, the focus of sanctions continued to be counter-proliferation. Between 2006 and 2010, as concerns with Iran's proliferation efforts again started to mount, American diplomacy turned to securing multilateral support through the United Nations, which passed a series of resolutions—including 1696 (2006), 1737 (2006), 1747 (2007), 1803 (2008), 1835 (2008) and 1929 (2010)—aimed at increasingly severe multilateral sanctions against Iran in response to its expanding enrichment efforts. Significantly, UN Security Council Resolution 1929 (2010) not only imposed an arms embargo on Iran but forbade Teheran to conduct nuclear-capable missile tests—a key component of any nuclear delivery programme.

In 2010 US sanctions against Iran escalated with the adoption of the Comprehensive Iran Sanctions, Accountability and Divestment Act, which had a focus on financial transactions. Extraterritorial sanctions were a key component of this Act, which targeted foreign banks undertaking business with designated Iranian banks as well as firms investing in Iran's energy sector. Any foreign bank that maintained a correspondent account with an Iranian bank was threatened with loss of its US banking relationships. The following year Congress passed further legislation imposing sanctions on the Iranian Central Bank and secondary sanctions on overseas banks and companies involved in importing Iranian oil.

In 2011 the Treasury identified Iran as a jurisdiction of primary money-laundering concern under Section 311 of the Patriot Act on the basis of Iran's pursuit of weapons of mass destruction and support of terrorism. The Treasury identified the entire Iranian financial sector, including Iran's Central Bank, private Iranian banks and branches, and subsidiaries of Iranian banks operating outside Iran as posing major finance risks for the global financial system. At the same time the Financial Crimes Enforcement Network (FinCEN) proposed imposing a special measure against Iran which would require US financial institutions to implement additional due diligence measures in order to prevent indirect access by Iranian banking institutions to US correspondent accounts.[34]

The US Iranian Transactions and Sanctions Regulations of 2012 (ITSR) made it illegal for US Persons to provide or receive goods or services from Iran.[35] In 2012 ITSR was also reinforced by Executive Order 13628 under the Iran Threat Reduction and Syria Human Rights Act, which extended the prohibition on US Persons transacting with Iran to foreign entities owned or controlled by US Persons. Also in 2012, the Iran Freedom and Counter-Proliferation Act included the threat of penalties against companies transacting with that country.[36]

The focus on banks remained an important theme throughout this period. Various regulations barred Iran from direct and indirect access to the American financial system. According to one report in 2013, the US Treasury approached 145 banks in 60 countries and convinced at least 80 non-US banks to cease dealing with Iranian banks.[37] Washington certainly acted aggressively against those breaching its sanctions regulations. In 2004 UBS was fined $100 million for unauthorised movement of dollars to Iran and other sanctioned countries; in 2005 ABN AMBRO paid an $80 million fine for improper dealings with the Iranian Bank Melli; in 2006 Credit Suisse paid a fine of $536 million for illegally processing Iranian transactions with American banks; in 2012 ING paid a fine of $619 million while Standard Chartered paid $340 million for Iranian financial sanctions breaches. In 2014 Clearstream Banking was fined $340 million while that same year the Bank of Moscow paid a $9.5 million fine for transferring money through the American financial system on behalf of Bank Melli. Heavier sanctions soon followed, and we have already mentioned the fines levied against BNP Paribas, HSBC and Commerzbank AG, which amounted to billions of dollars.

Initially at least the Europeans were hesitant to increase financial and trade sanctions upon Iran as they feared losing out on lucrative deals with its oil-rich economy. In July 2002 Iran was still able to sell a large amount of bonds to European banks. Indeed, between 2002 and 2005, while Washington was attempting to ratchet up counter-proliferation and counter-terrorism sanctions, the European Union sought greater economic co-operation with Iran by attempting to negotiate a trade and co-operation agreement which would have increased quotas and lowered tariffs between the two markets.[38]

The European Union had already for some time threatened to resist American attempts to impose secondary sanctions on EU entities. In 1996 the United States enacted the Iran-Libya Sanctions Act (ILSA) and the US Cuban Liberty and Democratic Solidarity Act (the Helms-Burton Act). These pieces of legislation contained sanctions targeting US subsidiaries incorporated overseas, including those interested in doing business with Cuba or transactions with the Iranian or Libyan energy sectors beyond a *de minimis* amount.

Resentment in Europe at the extraterritorial reach of both ILSA and the Helms-Burton Act and the need to protect European businesses against the increasingly long arm of American power resulted in the 1996 adoption of Regulation No. 2271/96—the 'Blocking Regulation'.[39] Should an entity subject to European Union jurisdiction be affected by extraterritorial sanctions, it was required not to comply with such a requirement whether actively or by way of omission or by way of an intermediary or a subsidiary. Entities were to notify the European Commission within thirty days if they became subject to any American extraterritorial measures. Such American sanctions would neither be recognised nor enforced within the European Union, though the Blocking Regulation allowed entities affected by the secondary sanctions to comply if non-compliance would have a serious negative effect on their interests or the interests of the European Union. EU persons would also be able to 'claw back' damages and legal costs if they were to be affected by defined extraterritorial sanctions. The Europeans also threatened to take the United States to the World Trade Organisation on this issue but in the end decided not to, as the hope emerged that Washington would itself not push too hard in implementing secondary sanctions.[40]

The first major European instance where the Blocking Regulation was used to combat US extraterritorial reach occurred in 2007 when Austrian authorities charged the country's fifth largest bank, BAWAG, with violating the Blocking Regulations when the bank—which had just been taken over by American investor Cerberus Capital—cancelled accounts belonging to about a hundred Cuban nationals. Austrian Foreign Minister Ursula Plassnik was quoted as saying that BAWAG had breached EU rules which forbade the implementation of US Cuban sanctions on European soil. 'U.S. law is not applicable in Austria. We are not the 51st [state] of the United States,' the minister stated.[41] At the same time she also referred to US objections against Austrian oil and gas group OMV's plans to produce gas in Iran, pointing out the slippery slope of acquiescing in US extraterritorial demands. Ultimately, the Austrian authorities dropped the case because the American purchaser obtained a licence from OFAC permitting the bank to retain the accounts. In this case, Washington appears to have backed down.[42]

For various reasons, including a convergence of perceptions as to the military intent behind Iran's nuclear programme, the EU sanctions regime increasingly began to align with that of the United States. So when Washington, especially from 2012, began expanding its sanctions toolkit targeting Iran, the Europeans did not take defensive measures, as they had in 1996. Rather they enacted a series of legislative steps which brought their sanctions policies generally, but not totally, into line with those of the United States.

The Europeans rapidly implemented the various Security Council resolutions in relation to Iran's nuclear activities and also imposed a wide set of autonomous financial and trade sanctions on the Iranian regime. These measures were consolidated in EU Regulation 267/2012 and enhanced in Council Decision 2012/635/CFSP. The measures included significant restrictions on financial dealings with Iran, including the freezing of assets belonging to the Central Bank of Iran and various other important Iranian commercial banks. There was a prohibition on Europeans investing in the Iranian oil and gas sector and providing any financing or technology in support. In addition, there was a ban on new Iranian bank branches opening within the European Union, the purchase by relevant entities of Iranian government bonds, and a requirement for the authorisation of transactions above €40,000.

In 2012 the European Union imposed an embargo on the import of Iranian oil as well as instituting a prohibition on the financing of insurance (the barring of insurance on Iranian oil shipments was a very effective method of reducing oil exports) and on the import and transport of Iranian oil and petrochemical products. At the same time, it imposed a ban on the export to Iran of certain technologies for the petrochemical sector, and there followed a freeze of Iranian Central Bank assets held within the European Union. In March 2012, the European Council instructed the Belgium-based SWIFT— the financial messaging network—to disconnect approximately 25 Iranian banks from its system. The Europeans also imposed a ban on all transactions between Iranian and member state banks. A blacklist of sanctioned Iranian entities and individuals was constantly expanded, and travel restrictions and asset freezes were imposed upon DPs.

By the time the JCPOA had been agreed in 2015, the Iranian economy was reeling under the weight of primarily American and European financial sanctions and trading bans. Following the ratcheting up of sanctions in 2012, Iran's economy contracted by 6.6 per cent. According to one 2015 estimate, Iran's

GDP was 15–20 per cent lower than it would have been without sanctions.[43] Petroleum products accounted for over 80 per cent of Iran's exports, and while in 2011 Iran exported approximately 2.5 million barrels a day (bpd), by 2014 exports of this key product were down to just over 1 million bpd.[44] Not surprisingly, between 2012 and 2015 Iran lost over $160 billion in oil revenues. A significant part of this loss was explained by Europe's reduction of its Iranian oil imports, and, indeed, between 2011 and 2014 EU bilateral trade with Iran declined from $28 billion to $12.8 billion. Nor could Iranian manufacturing escape the pinch of the increasingly aggressive targeted sanctions regimes. Production of automobiles for example fell by approximately 60 per cent from 2011 to 2013. The value of the rial declined over 50 per cent from January 2012 to January 2014. By 2014 unemployment was up to at least 20 per cent with inflation between 50 and 70 per cent.

During the JCPOA negotiations in exchange for restraints on its nuclear programme Iran was promised significant sanctions relief. The United States also indicated that following a deal it would lift the threat of secondary sanctions on non-American companies doing business with Iran.

The ultimate readiness of Teheran to put a hold on aspects of its nuclear programme for a set of fixed periods came no doubt in response to fears of an American or Israeli pre-emptive strike on its nuclear facilities. But it was certainly also a result of the promise by the United States to release tens of billions of dollars in frozen Iranian assets and the specific assurances of phased sanctions relief.[45]

In July 2015, following the signing of the JCPOA, the UN Security Council unanimously approved Resolution 2231. Essentially this nullified previous sanctions resolutions on Iran, though trade in conventional munitions would be restricted for five years and ballistic missile technology for eight years. There were various other so-called temporary 'sunset' clauses, that were causes of contention.

The United States, in turn, went on to release over $100 billion of frozen Iranian funds and arranged to cancel several previous executive orders containing sanctions aimed at that country. Over 400 Iranian individuals and companies were removed from the SDN List. US Persons were able to access licences to import certain foodstuffs and carpets from Iran. There was also permission to export aircraft and aircraft parts under a General Licence I.

However, the primary United States comprehensive embargo against Iran remained in place, including sanctions related to Iranian terrorism and human rights violations. These remaining sanctions applied to both US Persons and their owned or controlled non-US entities (so-called ITSR parties). Transactions with the government of Iran, parties remaining on the SDN list or parties usually resident in Iran, investments in Iran or sales to that country remained forbidden unless OFAC agreed to provide a general or specific licence. Facilitating non-ITSR parties in undertaking the work was also prohibited. Nor were transactions that used dollars or that were routed through the American banking system allowed.

What did alter was the form of extraterritorial sanctions, which had previously focused on non-US companies doing business with Iran. These were now suspended with the result that non-US companies could involve themselves in Iranian transactions including financial, oil, gas, shipping, insurance and automotive deals. Transactions could also be concluded with those 400 Iranian individuals and companies no longer on the SDN list. While US-controlled and -owned subsidiaries remained subject to ITSR, they could apply for a new General Licence H (subject to many restrictions, including no US Person facilitation) and also engage in many types of transactions with Iran.

Nevertheless, the reduction of extraterritorial sanctions was slightly illusory as non-US Persons still had to navigate around remaining US prohibitions, which could impact upon them dramatically. So Non-US Persons had to be very careful not to cause US Persons to contravene any of the remaining sanctions. There could be no dealings in US dollars or through the American banking system and no transactions with proliferators, terrorists and those Iranian parties remaining on the SDN list. While direct transactions with persons such as General Qassem Suleimani could presumably be avoided, the Iranian Revolutionary Guards, for example, remained heavily involved in a range of Iranian industries and companies, an involvement that could not be readily determined by the usual due diligence process. Global business continued to face significant challenges when it came to avoiding United States sanctions.

American steps to lift sanctions were matched by those of the Europeans. Following the signing of the JCPOA, the European Union lifted the oil embargo on Iran which it had imposed in 2012. Oil and gas and petrochemical product controls were raised as were investment and import restrictions on these sectors. Sanctions on shipping and gold and precious metals were removed. Restrictions on dealing with Iranian institutions, including the Central Bank, were lifted and hundreds of designated individuals and certain banks were de-listed. Fund transfer controls, correspondent banking restrictions, and restrictions on loans remained on certain designated parties including some banks and oil and gas companies. Restrictions also remained in force on trading in arms, nuclear and missile-related technologies, dual-use items, software for proliferation use, internal repression equipment, and raw and processed metals.

American–European tension was nevertheless built into the sanctions-related elements of the JCPOA because the remaining European restrictions were less onerous than those imposed upon American firms by United States authorities for whom the main sanctions regime still remained in place. Provided European companies avoided using dollars or the American banking clearing system or avoided involving US Persons, and provided they steered clear of Iranian individuals, institutions and items still subject to sanctions (admittedly not an easy task), they could in certain instances transact with Iran. In its eagerness to sign the deal the Obama administration appeared willing to tolerate a less-than-level playing field when it came to exploiting the opening up of Iranian markets.

In a statement on behalf of the European Union in May 2016, the willingness of the Europeans to push into Iranian markets was made clear: 'We will not stand in the way of permitted business activity with Iran, and we will not stand in the way of international firms or financial institutions engaging with Iran, as long as they follow all applicable laws.'[46] While American companies generally had to hold back (though there were exceptions, such as Boeing, Honeywell and Dover), their European counterparts such as Total, Airbus, Volkswagen, PSA and Renault, Siemens, Maersk and many others, backed by their governments, moved rapidly to enter agreements with Iran or consider expanding their investments in that country.

Nevertheless, restrictions on the use of dollars, the underlying threat of US secondary sanctions, the threat of snapback[47] and practical problems of doing business inside Iran inevitably served as a brake on European ambitions for expanded trade. By the time President Trump withdrew from the JCPOA in May 2018, Iran had benefited only partly from sanctions relief and certainly perceived itself as having received limited benefit. Although foreign investment in Iran had doubled since 2015, from $2 billion to $4 billion, this was a small amount in the context of an over 400 billion dollar economy.[48] Oil exports had certainly increased by 2017 to 2.1 million bpd but, as the Iranians recognised, this increase could be readily reversed by foreign action.

Much to the Europeans' discomfort the Trump administration, when reviewing the JCPOA, took a more negative view of Iranian nuclear intentions and a much broader assessment of Iranian behaviour, especially in relation to its activities in the region and its ballistic missile ambitions, activities which fell outside the confines of the JCPOA. In August 2017 President Trump signed into law the Countering America's Adversaries through Sanctions Act (CAATSA), which imposed new sanctions on Iran (as well as Russia and North Korea) and included within its remit expanded secondary sanctions.[49] In relation to Iran, CAATSA authorised the designation as SDNs of those persons who knowingly contributed to Iran's ballistic missile programme, who were IRGC officials, or who knowingly supplied weaponry to Iran, or who violated human rights.

When the withdrawal of the United States from the JCPOA occurred in May 2018 it was accompanied by a warning about the reimposition of secondary sanctions, a revocation of General Licence H and General Licence I, and a re-listing of SDNs that had previously been removed. The US withdrawal from the JCPOA was also accompanied with a timetable by which those companies that had begun to transact with Iran needed to unwind their deals. It set 90-day and 180-day wind-down periods for activities previously permitted under the JCPOA.

Certainly, one of the key consequences of the US withdrawal was the reimposition of American secondary sanctions and the impact this had on European businesses. After the 90-day wind-down period non-American businesses were unable to do deals with Iran's automotive sector; involve themselves in the acquisition by the government of Iran of dollar banknotes; directly or indirectly supply Iran with steel, coal, software and other goods; purchase Iranian sovereign debt, Iranian currency, or maintain accounts outside Iran denominated in rial; or trade with Iran in relation to precious metals. Following the 180-day wind-down restrictions were placed on transacting with Iran's shipping sector, petroleum and energy sectors, and the provision of insurance, reinsurance and underwriting services to Iran. Very significantly, on 4 November 2018, non-US financial institutions could not deal with the Central Bank of Iran and various other designated financial institutions, nor could they provide specialised financial messaging services to the Central Bank and Iranian financial institutions. To do any of these things would invite severe United States sanctions. In this way, Iran's financial sector was to be cut off from the global economy.

Once again tensions arose with the Europeans as OFAC guidance noted: 'Non-U.S., non-Iranian persons are advised to use these time periods to wind-down their activities with or involving Iran and that will become sanctionable at the end of the applicable wind-down period.'[50] The Europeans voiced strong opposition to the termination of the JCPOA and, in response, in August 2018 they updated the Blocking Regulations to take account of the new American actions. The aim now was to protect European businesses undertaking what the European Union considered to be 'legitimate' business with Iran from United States extraterritorial sanctions. In a joint statement by the EU, German, French and British foreign ministers it was announced unequivocally: 'We are determined to protect European economic operators engaged in legitimate business with Iran, in accordance with EU law and with UN Security Council resolution 2231 ... Preserving the nuclear deal with Iran is a matter of respecting international agreements and a matter of international security.'[51]

In 2018 and early 2019 no secondary sanctions were imposed on European companies. However, in sizing up their options, numbers of companies active in Iran preferred to ignore European Union remonstrations about bowing to American pressure by voting with their feet. Following the re-imposition of United States sanctions a variety of European companies soon began to exit Iran with some of them citing unspecified issues of commercial viability so as to exploit a carve out in the Blocking Regulation and thereby avoid triggering its implementation. Fear of United States secondary sanctions proved too powerful a threat. Other European efforts to contain American power in this space also proved difficult. Most significantly, attempts to create a special purpose vehicle which would participate in Iranian related transactions and thereby assume the risk on behalf of companies willing to engage with Teheran soon ran into a variety of problems not least overcoming institutional fears of Washington's response.[52]

Thus the case of Iranian sanctions clearly demonstrates the major effort that Western countries put into using financial instruments to coerce foreign states to move away from behaviour they considered unacceptable. At the core of this effort was a public–private coalition which placed international financial institutions on the front line of state efforts and required them to invest resources into helping conduct sanctions policy. As with the AML and CFT space, the firepower of the state was first turned on these domestic financial institutions (and in the case of the United States, foreign institutions as well) so as to ensure a strong and unified coalition capable of mounting an effective financial coercive policy. Where the realm of financial and other sanctions differed from the AML and CFT fields was that it was outwardly directed and touched more directly on foreign policy goals. As a result, because foreign policy objectives differed between the United States and European Union, tensions arose between states in implementing financial and other sanctions. This tension complicated the efforts of global financial institutions to help meet these foreign and security objectives while at the same time they sought to navigate through the minefield of state censure.

Brexit, Sanctions and UK–EU Tensions

The United Kingdom, while a member of the European Union, implemented that organisation's sanctions domestically pursuant to the European Communities Act 1972. It also implemented United Nations resolutions through domestic legislation or indirectly by way of EU regulations. There were also various autonomous UK sanctions in respect of terrorism. Brexit of course meant that these mechanisms needed replacement, and indeed the first piece of substantive Brexit legislation to receive royal assent was the Sanctions and Anti-Money Laundering Act 2018 (SAMLA),[53] which provided the British government with new powers so as to effectively operate its own sanctions regime.

SAMLA enables British authorities, through ministerial regulations, to impose financial sanctions, trade sanctions, and shipping, aircraft and travel restrictions which are 'appropriate' for purposes of implementing UN or autonomous sanctions, preventing terrorism, supporting national security, furthering British foreign policy objectives, promoting the resolution of armed conflicts or the protection of civilians in conflict zones, serving as a deterrent to violations of human rights, and contributing to efforts to stop the threat of weapons of mass destruction. SAMLA also provides the authorities with the power to add or remove persons and organisations from sanctions lists.[54] There are new Magnitsky-like powers—directly following the Skripal poisoning in Salisbury in 2018—to freeze and seize assets on the basis of 'gross violations of human rights', which are defined to mean torture and other cruel and degrading punishments conducted or approved by public officials.[55]

The 'appropriate' test for sanctions implementation is wider than the EU 'necessity' test, which means that ultimately the scope of an independent British sanctions regime may be broader than that of the European Union. Whether the United Kingdom would want to impose more intensive sanctions than its former EU partners and thereby surrender business advantage is another question entirely,[56] but what is clear is that it is quite possible that UK and EU sanctions regimes may over time, despite British government pronouncements on the advantages of co-ordination, begin to diverge. Should this happen, international sanctions regimes will inevitably be weakened.

For the United Kingdom private sector the more immediate implications of SAMLA are, unsurprisingly, an even greater compliance burden, which will only increase as sanctions regimes diverge. More specifically, SAMLA provides various challenges when it comes to designating persons to be subject to sanctions such as asset freezes. One important element of the Act is that it allows for the designation of persons not only by name but also by 'description'. From the start this possibility clearly alarmed business, which prior to the bill being enacted made it known that they feared the compliance uncertainty that would result from designations by description. The concern was that such designations (of both persons and entities) would inevitably be activated by the state to cover groups of persons in circumstances where those names were unknown to the authorities. The private sector was thus again being asked to do something that the state was unable to do itself. Demands upon the private sector's screening and reporting requirements will inevitably increase, with failure to comply becoming a criminal offence.

Guidance notes accompanying SAMLA have sought to give comfort to the private sector by stating that the power to designate by description would only be employed when it was not practicable for the minister to identify by name all the persons who fell within a particular description. In addition, the description would have to be precise enough for a reasonable observer to know whether a person fell within a particular description. The government also said that when exercising this power it would provide businesses with as much information as possible so as to enable them to comply with their obligations.[57]

How this will work in practice remains to be seen, but businesses are probably right to be wary. Even with the help of some of this promised information provided by the authorities (where available), they will now find that their usual screening of customers and other counterparties against the consolidated list of designated persons provides them with less confidence than ever that they are transacting within the law.

Export Controls and Embargoes

There were communication channels during the Cold War, and there was no obsession with Russophobia, which looks like genocide through sanctions.

Russian Foreign Minister Lavrov[58]

For over a year Ukrainian President Viktor Yanukovych promised that he would sign a trade deal with the European Union that would cement

Ukraine's increasing turn towards the West, but in November 2013 the President suspended talks with the Western powers. This soon led to violence in the streets of Kiev, which ultimately forced the President to flee the country. In March 2014 thousands of Russian soldiers in unmarked uniforms poured into Crimea and within two weeks Russia had annexed the peninsula—an action regarded by most of the international community as illegal. Shortly thereafter, separatists active in the eastern regions of Donetsk and Luhansk declared independence from Ukraine, and fighting escalated between these elements (backed by Russian forces) and Ukrainian forces. In May 2014 the new Ukrainian President, Petro Poroshenko, finally signed the EU Association Agreement and warned Russia to leave the country. Despite a ceasefire agreed to in September 2014, Russia continued to send additional troops into eastern Ukraine. While the Obama administration pondered whether to supply the Ukrainians with heavy weapons, Germany and France opposed this move, fearing such actions would merely worsen the already dire situation. In the face of deteriorating conditions in the east of Ukraine and a continued unwillingness to more forcefully support the Kiev government, the Europeans turned to tools they felt more comfortable with: a mixture of financial and trade sanctions coupled with military and technological embargoes.

The Russian–Ukrainian case study is a clear example of the practical overlap between sanctions and embargoes. Though implemented independently, both methodologies serve to support each other in the goals of coercing and constraining target behaviour as well as stigmatising and isolating regimes and individuals whose actions are considered unacceptable. Similarly, the case illustrates the challenges posed by American extraterritoriality, the necessity of public–private co-operation, and the difficulties faced by the private sector in adhering to state requirements especially when national policies differ.

While the primary aim of financial and other sanctions is to impact upon the activities of targets, export controls focus in the first instance on the cross-border movement of 'items' irrespective of targets. These items generally refer to the supply of strategic goods, services and technologies (GST) but in many instances have as their core focus weaponry and related finance and technologies—sometimes referred to as 'arms and related material'—and 'dual-use' goods. Decisions as to which GST will be subject to control are determined, like sanctions, on the basis of national security, human rights, proliferation and terrorist concerns. When general export controls are linked specifically to targets, they manifest in the form of embargoes and often form part of a general sanctions regime. Exporting to an embargoed destination may still be possible if the relevant authority issues the correct licence. Failure to comply with embargoes, licensing or end-use control requirements may result in criminal penalties and imprisonment.

Viewed from the perspective of a company seeking to export an item and at the same time comply with export controls and any embargoes, the term 'export' is broadly construed and covers physical transfers and hand-carried items (such as computers containing software), and intangible transfers (such as email and electronic downloads). Moreover, for the purpose of considering controls, intra-group transfers are also relevant, as are transfers within a country or even between third countries. Indeed, for an export to have taken place no sale is actually required.

In this context, exporters will need to take into account several key questions. These questions include the following. Is the product they intend to export subject to export controls? Could the product be of military or other sensitive use? What is the purchaser intending to use the product for and does the company understand why the purchaser wants the product? Is the end-user different from the immediate purchaser? Where is the item being sent to and is the jurisdiction subject to sanctions or embargoes? Is the jurisdiction to which the item is being dispatched a diversion risk? What is the method by which the company will ultimately be paid and will the sanctions impact upon the receipt of payment? And what is the policy of the home jurisdiction and the company itself in relation to the territory of the purchaser or end-user?

As with sanctions regimes, the sources of export controls and embargoes may be supranational (United Nations), regional (for example, the European Union) or national. It was the European Union and its member states that took the lead in export controls aimed at Russia because of its behaviour in Ukraine, and it is to the European Union that we shall first turn.

EU export controls and embargoes apply to persons who are subject to EU jurisdiction, irrespective of the origin of the items. The centrepieces of the European approach are the EU Common Military List,[59] and the EU Dual-Use List and its annexes (EU Regulation 428/2009), which as part of a regulation has direct applicability in member states.[60]

The EU Common Military List sets out the list of weaponry in relation to which members need to control the export. Goods and technologies are classified as military goods if they are designed specifically for military use, such as small arms and armoured vehicles.[61] Items are considered to be dual-use when they can be employed for civil and military purposes. Annex 1 of the EU

Dual-Use List is divided into categories, with each category covering a different technology. So, for example, Category 0 covers nuclear materials, facilities and equipment; Category 4 covers computers; and Category 9 includes aerospace and propulsion.[62] No licence is required for transfers within the European Union of most of these items, but they are controlled if exported from the European Union, and some jurisdictions such as the United Kingdom do require licences for intra-EU transfer if this is followed by an export out of the European Union (a so-called indirect export). Companies will need to make clear in their documentation of Annex 1 items that those items will be controlled if exported out of the European Union. Licences will also need to be secured if for example a company wishes to export out of the European Union certain items listed in Annex IV, such as explosives and pressure sensors. The European Union does not include deemed export controls or re-export controls such as are found within the American system, though specific EU licence conditions may still be applicable.[63]

The EU export control regime is based on national licensing and enforcement capabilities. In the United Kingdom—which at the time of writing is still part of the European Union and will effectively be aligned with that organisation in terms of export controls for some time to come—the UK Strategic Export Control Lists can be found in the United Kingdom's consolidated list of strategic military and dual-use items that require export authorisation (the 'Consolidated List'). The Consolidated List is made up of several lists: (i) the EU Dual-Use List; (ii) the EU Human Rights List (consisting of, for example, items that could be used for torture); (iii) the UK Dual-Use List,[64] which expands the EU variant; (iv) the UK Military List;[65] (v) the UK Security and Human Rights List; and (vi) the UK Radioactive Sources List. If an item a company wishes to export is found on one of these lists, it is controlled (or 'rated'), and British exporters may need to secure a licence for their export from the Export Control Joint Unit (ECJU), which is part of the Department for International Trade and which deals with the export of strategic goods.[66] Companies need to check these lists as they are regularly updated, and they need to be aware that there is no rating validity period during which they will be protected.

The UK Military List, which comprises military goods, technology and software, is based upon the EU Common Military List with various British additions. The focus is on military hardware and items 'specifically designed or modified for military use'. Different types are given an ML designation; so, for example, bombs and torpedoes are designated ML 4, military vehicles are designated ML 6, and related technology ML 22. States within the European Union have their own lists but they broadly mirror those of the United Kingdom (such as guns, armoured vehicles, missiles, naval vessels and fighter jets) though differences may emerge when it comes to components. All items on the UK Military List are controlled for exports outside the United Kingdom, including to EU states. Significantly, the fact that products are employed by the military—e.g. umbrellas—does not mean that item is controlled. Conversely, military helmets used by construction workers would be construed as controlled.

European Union countries and the United Kingdom follow the EU Dual-Use List supplemented by their own dual-use lists. Also, some EU states seek to control the 'trafficking and brokering'—that is, supply and delivery activities, including negotiating, referring and providing finance—of military goods (for Britain, it is those goods on the UK Military List) between third countries.[67] So, for example, a person cannot seek to circumvent export controls by shifting the business location outside the relevant jurisdiction.

What are the most important licences for which European and UK exporters need to consider applying? EU General Export Authorisations (GEAs), known as 'Union Licences', license dual-use items and are valid in all member states and may be used by any exporter established in the European Union. There are six EU GEAs:

- EU001—exports to Australia, Canada, Japan, New Zealand, Norway, Switzerland (including Liechtenstein) and the United States;
- EU002—export of certain dual-use items to certain destinations;
- EU003—export after repair/replacement;
- EU004—temporary export for an exhibition or fair;
- EU005—telecommunications;
- EU006—chemicals.[68]

In the United Kingdom, Open General Export Licences (OGEL) are pre-published export licences issued by the ECJU. They are a flexible type of strategic export licence in that, provided all pre-set conditions can be met, they permit the export of specified controlled items to specific destinations.[69] Also in the United Kingdom, a Standard Individual Export Licence (SIEL) is the most common type of strategic export control licence. It gives an exporter permission to export specific items to specific destinations. The exports must be addressed to a stated person or company, and there must also be an identification of the end-user. The goods have to be of the same value and quantity as described on the licence. Companies will apply for a SIEL if they cannot meet all the required terms and conditions of an OGEL.[70] Finally, an Open Individual Export Licence (OIEL) is a form of concessionary licence available to individual exporters who have a track record in applying for export licences or who can demonstrate a business case for such a licence. OIELs cover multiple shipments of specific controlled goods to named destinations. They may name the consignees or end-users of the goods, unlike SIELs, which always name these parties.[71] Companies will decide upon the relevant licence application depending upon the business conducted, the item to be exported and the target destination.[72]

Even if a company is confident that the products it intends to export are not located on any of the control lists, it still needs to take into account the issue of end-use controls. End-use restrictions are relevant, independent of controls on the export or re-export of items. Here a licence may be required if there are concerns about the end use of the items in question. Two areas are of importance: weapons of mass destruction (WMD) and military end use.

If an EU or UK exporter knows, or suspects, or has been informed[73] by competent authorities that the product is intended for WMD end use outside the European Union—WMD end use being widely construed as including development, production, maintenance, storage and dissemination—then a licence will be required.[74] Of course, the question arises as to how a company would know or suspect that information unless explicitly informed. Suggested 'red flags' here vary from the explicit, such as the end-user appearing on a jurisdiction's WMD watch list, to the more general, such as failure to respond to routine technical questions, unusual shipping requests and unusually favourable payment terms. Whether all companies would be sensitive to some of these factors in an end-user context is a questionable point.

Military end-use controls are essentially a catch-all device that aims to sweep up items that may not be on any list but could be put to malign ends. In the United Kingdom a licence will be required if the exporter is aware or has been informed by the competent authorities that (i) dual-use items are or may be intended for use in military equipment in embargoed destinations or (ii) dual-use items may be intended or used as parts of military goods which were illegally obtained from the United Kingdom, irrespective of the actual destinations. Companies will need to screen both against the customer and the destination for end use.[75]

In 2018, key UK destinations subject to embargoes included Armenia, Azerbaijan, Belarus, Burma, Central African Republic, the Democratic

Republic of Congo, Eritrea, Iran, Iraq, Lebanon, Libya, North Korea, Russia, Somalia, South Sudan, Sudan, Syria, Venezuela and Zimbabwe.

Companies in Europe and all over the world cannot avoid taking into account US export controls, as the relevant jurisdictional reach is wide and touches upon foreign incorporated subsidiaries owned or controlled by US entities. American jurisdiction will also apply to exports, re-exports and in-country transfers of what is termed 'US origin items'. It will thus cover exports from the United States, re-exports from other countries of not only 100 per cent US-origin items but also exports of non-US-origin items if they exceed a defined minimum of American content. Furthermore, US jurisdiction will extend to situations where the export involves no US content but the items are based upon US technology. Companies need to bear in mind here that American jurisdiction will follow the item outside the United States to all destinations, not only those countries subject to sanctions.

The two relevant US regulations which apply respectively to weaponry and dual-use embargoed items are the International Traffic in Arms Regulations (ITAR), which fall under the Arms Export Control Act 1976, and Export Administration Regulations (EAR), which come under the Export Administration Act 1979. ITAR has a primarily military focus and covers military items and defence articles. EAR deals with dual-use items and regulates those goods and technology which have been designed for commercial purposes but which could be put to military use. Here, too, a licence will be required for export.

Linked to these regulations are two lists: the US Munitions List (USML), which is connected to ITAR, and the Commerce Control List (CCL), which is connected to EAR.[76] Specifically, ITAR jurisdiction covers items on the USML, those items that provide equivalent performance capabilities of a defence article that appears on the USML, various services that are provided in support of defence articles, and items that are able to provide a military or intelligence advantage to those employing them and therefore warrant control.[77] Those items not covered under ITAR will fall within EAR and the CCL. These include non-military commercial items, technology and software, as well as dual-use items. Each entry in the CCL is provided with an export control classification number (ECCN) and is described by technical parameters.[78] Items that do not appear within the CCL are termed EAR 99 items and, as long as there are no sanctions or end-use considerations, they may be exported without a licence.

Licensing policy is determined by considerations of national security, foreign policy or short supply, and is in turn informed by considerations of regional stability, proliferation, counter-terrorism and crime control purposes as well as adherence by the United States to the various multilateral and non-proliferation control regimes. The Directorate of Defense Trade Controls within the Department of State is responsible for enforcing ITAR items while the Department of Commerce's Bureau of Industry and Security (BIS) (where 'industry and security intersect') licensing authority[79] is responsible for EAR's CCL list. Significantly, BIS controls will, as noted above, impact on Non-US Persons who export or re-export items subject to EAR.

The American system imposes end-use controls in respect of WMD and missile concerns and military end-use controls for particular ECCNs in relation to certain targets (for example, Russia).

As with the US approach to money laundering, terrorist financing and sanctions violations, companies (wherever they are located) have to be aware of American penalties because export control violation penalties can be substantial. Breach of ITAR may incur civil fines of $500,000 per violation and criminal penalties may equal $1,000,000 and ten years' imprisonment per violation. Breach of EAR can amount to $250,000 per violation and criminal penalties up to $1,000,000 and twenty years' imprisonment per violation.

Sanctions and Embargoes Targeting Russia

It was with these sanctions and embargo tools that the United States and the Europeans began fashioning a response to Russian activities within Ukraine. Russia posed a very different challenge to the West from that of Iran or, as we shall see in the next chapter, North Korea. This stems from the fact that Russia boasts a sizeable and sophisticated economy which has global reach but according to critics of the country bears many of the characteristics of a sophisticated kleptocracy with President Putin and his key associates at its core. The aims of both the US and EU sanctions and embargo regimes was ultimately to force Moscow to withdraw its forces from eastern Ukraine and Crimea and to return the position to the status quo ante. To this end the United States stated that it would work with the European Union to increase costs to Russia until it 'abides by its international obligations and returns its military forces to their original bases and respects Ukraine's sovereignty and territorial integrity'.[80]

Key relevant US legislation here includes Executive Order 13662 (2014) and also the Countering America's Adversaries through Sanctions Act (CAATSA 2017), which covers secondary sanctions within its scope.[81]

Since the beginning of 2014 Washington has imposed a complete embargo on Crimea. When in March of that year the United States began ratcheting up related pressures on Russia the method by which this was conducted was by imposing so-called sectorial sanctions which relied upon targeted financial and trade sanctions supported by embargoes directed at key sectors of the Russian economy as well as President Putin's coterie.

OFAC enforced sectoral sanctions by restricting financial and other dealings undertaken by persons subject to US jurisdiction with state-controlled companies active in Russia's financial sector and viewed as key to Putin's financial power. These included Bank Rossiya (sometimes referred to as the personal bank of Putin), the Volga Group and companies and banks such as Gazprombank, Vnesheconombank (VEB), Rosneft and SMP Bank. US Persons were prohibited from undertaking certain types of financial transactions, such as providing new debt having maturities beyond thirty or ninety days, with entities such as Rostec, Sberbank and Rosoboronexport.

Also targeted were individuals and entities active in Russia's energy sector and financial sanctions were accompanied by embargoes on specified oil-related exports and services. American entities and individuals were restricted in providing goods, services and technologies that could support the exploration or production of Arctic offshore, deep-water or shale projects that produced oil. Significantly, the Ukraine Freedom Support Act took on an extraterritorial dimension when it placed sanctions not only on such activities but also on foreign financial institutions that funded these energy development projects or transacted with persons subject to relevant sanctions.

The Russian defence sector unsurprisingly became subject to this sectoral drive. American authorities prohibited export licences for military and dual-use goods as defined in the US Munitions List and the CCL to Russia. They also targeted for sanctions Almaz-Antey, a major Russian defence company, the arms exporter Rosoboronexport and other companies engaged in the transfer of weapons to Ukraine as well as various important Russian defence entities.

In relation to Crimea, OFAC implemented a comprehensive sanctions regime while BIS imposed a comprehensive export and re-export ban. American authorities made efforts to align OFAC sectoral sanctions with BIS export controls to ensure effectiveness. An EAR Entity List was set up and included various entities which appeared on OFAC's SDN list and which are subject to licensing,[82] such as the Russian Federal Security Services and, within Crimea, FAU Glavgosekspertiza Rossi.

The United States also sought to block assets of selected individuals and entities which had, in the opinion of the American authorities, undermined the stability of Ukraine, been involved in the annexation of Crimea, or had since conducted business in that region. Many of the individuals targeted included politicians and business persons with close associations to the Kremlin. For example, US authorities froze assets within the United States and prohibited transactions with Igor Sechin, head of Rosneft, Sergei Chemezov, director of Rostec, and Yuri Kovalchuk and Nikolai Shamalov, major shareholders in Bank Rossiya—all part of President Putin's personal and institutional power circle. These sanctions were complemented with those targeting individuals involved in human rights abuses and corruption in Ukraine. In April 2018 OFAC designated 7 Russian oligarchs, 12 of their companies and 17 government officials together with a bank and weapons trading company to the SDN list.

American sanctions then intensified further in response to the Novichok chemical attacks in Salisbury, England, in 2018. In August of that year new sanctions were imposed on Russia under the Chemical and Biological Weapons Control and Warfare Elimination Act 1991. Under these sanctions, efforts by American companies to secure export licences for items with a potential security purpose would be automatically denied. Exporters could try to demonstrate that the items were for legitimate civilian purposes, but that was expected to be a difficult case to make.[83] Since 2014 the US has targeted approximately 600 Russian-related entities and individuals with sanctions and embargoes.[84]

As we have noted, CAATSA 2017 reinforced secondary sanctions which impacted upon Non-US Persons. Not only were Russians targeted but so were those non-American businesses that were, in the opinion of the United States, supporting the Russian targets. Specifically, sanctions were applied against not only US Persons but also non-American financial institutions that knowingly undertook significant financial transactions with SDNs. There were also sanctions imposed on Non-US Persons who made significant investments in 'special Russian crude oil projects'. Similarly, persons facilitating the evasion of sanctions were targeted as were those dealing with the Russian intelligence and defence sectors. Sanctions also targeted Russian government officials as well as their family members and close associates involved in 'significant corruption'. Also targeted were those involved in serious human rights abuses in areas controlled by the Russian government. In September 2018 OFAC and the State Department demonstrated their extraterritorial reach when relying on CAATSA they sanctioned a Chinese organisation which was part of the Chinese military industrial complex for transacting with the Russian defence and intelligence sector.

Putting aside this threat of sanctions applicable to Non-US Persons, for Washington's approach to have gained significant traction against Russia it would also have required voluntary European co-operation. Russia's trade contracts with Europe far outstripped those with the United States. In 2015 almost 50 per cent of Russian exports went to the European Union as compared to less than 3 per cent to the United States. Moreover, almost 40 per cent of Russian imports came from the European Union as compared to approximately 6 per cent from the United States. In financial terms, by some estimates, nearly 75 per cent of Russia's overseas bank loans originated from European banks. Conversely, Russia remained a major supplier of oil and natural gas to the European Union, a factor which, in theory at least, mitigated some of Moscow's vulnerability to European sanctions[85]—a situation which President Trump severely criticised.

The EU approach has not been too dissimilar to that of the United States and efforts have been made to co-ordinate actions, but Washington's sanctions have in their totality been more comprehensive than those of the Europeans. American export controls applied extraterritorially, which meant that European organisations were required to review both sets of sanctions and embargoes. Not all companies and individuals targeted are the same.[86] For example, organisations such as Gazprom, Novatek and Lukoil have been targeted only by the United States. Furthermore, the tests for determining ownership of DPs and SDNs were different and could lead to varying results. European companies need to run both tests if they are to be confident that they are avoiding contravening American sanctions. The European Union also introduced restricted grandfathering provisions, which means that old deals entered into before sanctions were implemented need not necessarily all be unravelled—though European companies would be well advised to be extremely careful here as this could potentially raise problems with the American authorities.

By 2018 the EU sanctions regime, which like that of the United States had begun in March 2014, was targeting over 150 persons and nearly 40 entities with asset freezes and travel bans—or, as the Europeans express it, with 'restrictive measures'.[87] From the summer of 2014, the European Union began imposing sectoral sanctions and embargoes upon Russian interests. When by the end of 2015 Russia failed to implement the Minsk Protocol aimed at curtailing the fighting in Ukraine, these sectoral financial sanctions and specified embargoes were extended. The main elements of the European approach included a mixture of sanctions and embargoes and restricted access for certain Russian companies and banks to EU primary and secondary capital markets. An arms import and export ban was also imposed in respect of items appearing on the EU Common Military List (with various exceptions) and an export ban on items appearing on the EU Dual-Use List. In addition there is a restriction on the access by Russians to certain technologies and services that can be employed for oil exploration and production, including those capable of producing oil from resources located within shale formations through hydraulic fracturing. Clearly, the intent is for the European approach to parallel that of the United States.[88]

Restrictions on financial co-operation have been emphasised with the result that the European Investment Bank has suspended all new financing operations with Russia as has the European Bank for Reconstruction and Development, while various bilateral and regional financial co-operation programmes have also been suspended. EU organisations and nationals can no longer buy or sell new bonds, equity or other types of financial instruments from designated Russian banks, energy companies and defence companies, including their subsidiaries. Loans with maturities exceeding 30 days cannot be provided to the designated entities. All this has been accompanied by the suspension of the G8, which reconvened without Russia as the G7.

While the United States has imposed a comprehensive embargo on Crimea, the Europeans have imposed extensive restrictions. The Europeans focus specifically on limiting economic and financial relations with the Russian-occupied Crimea and Sevastopol. It imposes an import ban on goods from that region together with restrictions on trade and investment in various economic sectors and infrastructure projects. Companies within the European Union cannot purchase real estate or entities in Crimea and Sevastopol. In addition, an export ban has been imposed on goods services and technology for the transport, telecommunications and energy sectors and for the exploration of mineral resources, oil and gas. Neither can technical assistance, construction, engineering or brokering services be provided for those same sectors. Tourism services have been halted.[89] Asset freezes have been imposed upon individuals who have benefited from the transfer of property ownership contrary to Ukrainian law.

Every member of the European Union has undertaken national measures to implement and reinforce the organisation's decisions. In the United Kingdom, companies still need to be aware that sectoral sanctions imposed on Russia do not appear on the UK Consolidated List but rather on a separate Capital Markets List dealing with the Russian–Ukrainian situation.[90] These are required to be separately reviewed together with US sanctions and embargoes so as to avoid any infringements of both European and American measures. Each EU member has a somewhat different approach towards implementation but the focus is sectoral and there is a continuing requirement to avoid contravening US policies.

Since March 2014 Russian-targeted sanctions and embargoes have been constantly updated and existing ones prolonged. These sanctions measures have been complemented by the European Union granting various trade preferences to Ukraine, which have had the effect of permitting several industrial and agricultural Ukrainian goods better access to European Union markets. Concomitantly, there has been a lessening of restrictions on US military sales to the Ukrainian armed forces.[91]

Soon after the implementation of sanctions in 2014 and certainly by 2015, Russia began feeling the effects. In 2015 the rouble underwent a rapid depreciation of over 50 per cent against the dollar, and growth, which had already begun to reduce in 2014, continued along this path during the following year. Inflation increased from a quite high 6.8 per cent in 2013 to 15.5 per cent in 2015 with the budget deficit widening to 3.3 per cent in 2015. Reserves began to plummet dramatically from $500 billion at the beginning of 2014 to $368 billion at the end of 2015. Net private capital outflows began accelerating and in 2014 totalled $152 billion as compared to $61 billion in 2013. This would certainly seem to indicate a high level of impact, though it is very difficult to tease out the effects of sanctions from the then concurrent drop in the oil price, oil being a major source of Russian export revenue.[92]

While access to foreign capital remains severely restricted and foreign bank loans to Russia fell from $225 billion at the end of 2013 to $103 billion by the third quarter of 2016, the Russian economy then stabilised somewhat, with inflation falling to 7.2 per cent and the economy continuing to contract but at a somewhat slower rate. The rouble also began to strengthen and private sector capital outflows slowed to approximately $15 billion.[93] Russia resumed bond sales and net inward foreign direct investment started to pick up. This partly reflected higher oil prices, flexible exchange rates and still-sizeable foreign exchange reserves.

However, reduction of economic output has never been the ultimate purpose of Russian targeted sanctions and embargoes. That goal was always the withdrawal of Russian forces from Ukrainian territory. But by 2019, despite all these clear economic pressures, Russia showed little indication of reconsidering its military positions. Indeed, if we broaden our perspective and take into account Russia's deployment of forces to Syria as well as its increasing focus on strategic weaponry, sanctions and embargoes appear to have had little effect—at least so far—on Russian foreign policies.

Among her first remarks as US Ambassador to the United Nations, Nikki Hailey stated in early 2017 that 'Crimea-related sanctions will remain in place until Russia returns control of the peninsula to Ukraine'.[94] The stand-off seems to still have a long way to run.

Athens versus Melos

We began this chapter by referring to the Iran–Iraq War of the 1980s. Somewhat earlier, in the sixteenth year of the Peloponnesian War (431–404 BC) the Athenians descended upon the small island of Melos with more than 30 ships and accompanied by 1,600 heavy infantry, 300 archers and 20 mounted archers together with their allies. The Melians were refusing to join the Delian League and greatly frustrated the powerful Athenians, who, intent upon strengthening their coalition against Sparta, threatened Melos with severe retribution if they did not bend to Athenian demands. In what became known as the Melian Dialogues the leaders of both parties sought to convince each other of the strength of their respective positions. While the Melians couched their arguments in terms of what they considered the essential fairness, justness and righteousness of their cause, these sentiments were harshly brushed aside by the Athenians. 'For ourselves,' said the Athenians, 'we shall not trouble you with specious pretences ... and make a long speech which will not be believed ... since you know as well as we do that right, as the world goes, is only in question between equals in power, while the strong do what they can and the weak suffer what they must.'[95]

It is through this Athenian perspective that it is best to comprehend US extraterritoriality when it comes to sanctions and embargoes. The necessity of domestic coalition-building between the public and private sectors is complemented in this context of sanctions and embargoes by the necessity of coalition-building at the international level. The end to which sanctions and embargoes are put differ from AML and CFT policies in that they aim at coercing external behaviour and not in the first instance defending the viability of domestic systems against malicious intrusion. Thus the support of other states, and the private sectors in those states, is critical in fashioning effective policies. This is why US extraterritorial sanctions are so central to American sanctions policy.

To this end Washington is willing to flex its diplomatic and economic muscles to bring other states, together with banks and companies in such overseas jurisdictions, into line with American foreign and security policies, be that in relation to Iran, Russia or elsewhere. The Europeans may carp at this brutal exercise of force and bristle at its callous disregard for their sovereignty, but in the final analysis they will inevitably bend to US will. As trade and compliance lawyer Ross Denton has noted in respect of EU threats about reinvigorating the Blocking Regulation in the wake of the American withdrawal from the JCPOA: 'The US rock is far more compelling than the EU hard place, and very few EU businesses will rely on the Blocking Regulation to guarantee their ability to keep doing business in the US and Iran ... The US financial system is now so important to global and EU businesses that it cannot easily be avoided.'[96]

In evidence in 2017 before the House of Lords on Brexit and sanctions policy, the concern of international business about US sanctions was made absolutely clear. Matthew Findlay, Deputy Head of International Organisations in the Foreign and Commonwealth Office, noted that for businesses the biggest concern was US sanctions 'because they know that OFAC is very aggressive, the fines are very high and the public censure is very serious' and that EU action was not comparable to this. Dr Clara Portela of Singapore Management University concurred with this assessment when she stated: 'Basically [companies] are afraid of OFAC.' These analyses were supported by Francesco Giumelli of the University of Groningen in the Netherlands when he commented that, compared to American sanctions, 'companies do not care too much about [the sanctions of] the EU, Switzerland and Norway.'[97]

Given the size of American fines, the aggressiveness of OFAC, and the declaratory policy of the Trump administration, these are not groundless concerns. Consequently, in a choice between whether to avoid US or EU sanctions and embargo policies, most European and global businesses would be best advised to take great care not to incur American censure. And indeed as noted, by early 2019, despite the Blocking Regulation being updated and in force, many European businesses decided that their interests were best served by reconsidering their positions in Iran.

The Melians were ultimately not able to view the situation with such bracing clarity. Following a devastating siege they surrendered to the much stronger Athenians, but the vengeful Athenians had all the Melian men put to death and the women and children sold into slavery.

7

COUNTER-PROLIFERATION FINANCE

We have built the infrastructure needed to combat the financing of proliferation: with criminal laws and investigative powers; due diligence and suspicious transaction reporting by financial institutions; and transparency requirements regarding the control of corporate vehicles and legal arrangements.

FATF President Juan Manuel Vega-Serrano, December 2016

Top Cape Technology and the Pakistan Connection

Top Cape Technology was an import-export business based in Cape Town, South Africa. In 2003 the company's principal made contact through a broker in Secaucus, New Jersey, with a private manufacturer in Salem, Massachusetts, for the import to South Africa of approximately 200 triggered spark gaps. These high-speed electrical switches are often used in hospitals to power lithotripters which can break apart kidney stones, and indeed Top Cape Technology informed its American contact that the end-user would be Baragwanath Hospital in Soweto, the largest public hospital on the African continent. The only problem was that the hospital had not sought to purchase the spark gaps and had never heard of the Cape Town import-export company.

In fact, spark gaps are a type of dual-use item that also serve as detonators whose synchronised electronic pulses can trigger nuclear devices. And, as United States court documents would later attest, the intended end-user was not a Soweto hospital but a business in Islamabad, Pakistan, with links to the Pakistani military and various militant groups in the Kashmir. US law requires

215

an export licence for spark gaps when selling them to countries of nuclear proliferation concern, including Pakistan, though not South Africa, which had surrendered its nuclear weapons in the 1990s. Once the Cape Town company had received the spark gaps, which arrived in three shipments each of 66–7 items, they were quickly flown first to Dubai and then on to Islamabad.

Underpinning the proliferation transaction was a financial transaction. Front companies and false billing records were employed to disguise the purpose and destination of the goods while the shipment was financed by a letter of credit opened at the National Bank of Pakistan with the Standard Bank of South Africa. A letter of credit is a letter from a bank guaranteeing that a purchaser's payment to a seller will be received on time and for the correct amount. If the purchaser does not pay, the bank issuing the letter of credit will need to cover the full or remaining amount of the purchase. Top Cape Technology was thus confident that it would ultimately be paid for the technology transfer in which it was involved.

For Top Cape Technology to have been paid, a representative would have needed to take the letter of credit to Standard Bank together with documentation proving that the contract has been properly completed, and that the correct items had been delivered to the correct consignee by a specified date. Standard Bank would then have paid Top Cape Technology the agreed sum and recouped the money from the National Bank of Pakistan, which would have received the funds deposited into a local account by the Pakistani business. When Top Cape Technology shipped the spark gaps via Dubai to Pakistan, the National Bank of Pakistan was named as the consignee.

The spark gaps that the company sent to Pakistan were inoperable. They had not been filled with gas and their gas in-take lines had been pinched closed. The American authorities had been tipped off about the deal from an anonymous South African source and the Salem-based exporter had ensured that the items were disabled by the time they left the United States.[1] The director of Top Cape Technology was later arrested in Denver for violating the Export Administration Act and imprisoned.

Proliferation as a Business Cycle

Sanctions regimes and embargoes certainly target proliferators by seeking to impose general economic hardship on the proliferator and forcing the latter off the path of proliferation. The Iran case provides a prime example of such a strategy, and North Korea too has been targeted in this manner. But a subset of financial tools focuses on the financial disruption of the actual proliferation process itself.

Underlying this approach is the recognition articulated by Juan Zarate: 'By looking at the threat as a business cycle—which originates with the source of nuclear material and proceeds with the smugglers, facilitators, and end users—we can find multiple points at which to influence and interrupt the supply chain.'[2] Proliferators acquire financial obligations at various points of the proliferation process, which they need ultimately to fulfil. Payments to manufacturers, letters of credit, shipping fees, insurance and broker fees are but some of these financial elements that undergird the proliferation process and may at the same time also be vulnerable to identification and interdiction.

Counter-proliferation finance is based on the view that proliferation finance is a distinct form of financial activity that by its illegal nature also presents a threat to the global financial system. This financial threat requires bespoke financial solutions. Furthermore, counter-proliferation finance represents another weapon that can be used to support broader foreign and security policy directed at terrorist groups and rogue states.

In general terms, counter-proliferation finance has the following main goals, some which overlap with other areas of finance security and some of which are unique. These include:

- To defend the global financial system from proliferation financing activities;
- To prohibit the provision of a broad range of financial services in support of weapons of mass destruction (WMD) proliferation programmes and delivery systems (such as ballistic missiles);
- To add to the financial challenges facing proliferators and thereby hamper their efforts;
- To prevent financing of particular shipments in support of WMD proliferation;
- To contribute to stopping and seizing the flow of funds in relation to WMD proliferation;
- To support financial investigatory services in relation to proliferation; and
- To contribute to both the identification and disruption of proliferation networks.

As will be readily acknowledged, given the technical and political aspects associated with WMD proliferation, some of these counter-proliferation finance objectives are reasonably ambitious when viewed from the perspective of private sector institutions.

In this demanding task, once again, the private sector—financial institutions, export agencies and other businesses—have been brought in to play a significant role in stopping the spread of weapons of mass destruction. Banks and corporates are now faced with an additional compliance challenge to their other demanding compliance requirements, but one that they are singularly ill equipped to perform. They can be asked to focus on lists or companies and individuals provided to them by authorities which they mechanically check—a task which is not without its problems but is arguably manageable. They can also be asked to focus on the items and technologies themselves, an undertaking that in some instances requires a level of expertise beyond that of the average compliance office, let alone that of bankers, other corporate executives or even import-export professionals.

And, most ambitiously of all, these parties are also required to be 'vigilant' in respect of the signatures of activities indicating proliferation, a role perhaps better suited to intelligence officers than finance professionals. This endeavour is made even more complicated by the fact that the focus of their efforts is on preventing a set of actions that are often indistinguishable from legitimate activities.

Terrorists and Proliferating States

The Financial Action Task Force (FATF) approach to counter-proliferation financing—which currently forms the centrepiece of the global effort in this field—has to be placed within the context of UN policy towards this same goal. The UN approach preceded the steps initiated by FATF and is given expression in resolutions that are adopted by the UN Security Council under Chapter VII of the United Nations Charter and that are binding upon member states.

First, there are general broad-based provisions under Resolution 1540 (2004) and successor resolutions. These emerged in the wake of the 9/11 terrorist attacks and required member states to adopt measures to prohibit proliferation activities specifically by non-state actors, including prohibiting the provision of funds and financing services that could support WMD proliferation as well as their means of delivery.

In the early 2000s there was a clear and emerging concern that al-Qaeda was interested in WMD, especially in acquiring radiological weaponry. In a 1998 interview with *Time* Magazine, Osama bin Laden asserted that obtaining weapons of any type was demanded by his belief system and that

'Acquiring weapons for the defence of Muslims is a religious duty. If I have indeed acquired these weapons, then I thank God for enabling me to do so.'[3] It may seem slightly far-fetched now but at the time there was a stream of reports that al-Qaeda was acquiring WMD material—reports that could not be discounted in their entirety given that they coincided with and sometimes were based upon other accounts of smuggling of nuclear weapons-related material out of the former Soviet Union. These concerns were reinforced by evidence emerging from court cases in the United States involving captured al-Qaeda personnel such as Jamal al-Fadal and Ahmad Ressam, and also intelligence generated during the interrogation of the al-Qaeda operative Abu Zubaydah. Several scientists who may have been in contact with Osama bin Laden himself in connection with radiological weapons were also arrested by Pakistani authorities during this period. Finally, in 2003, there was a series of anthrax attacks in the United States which, while not linked to al-Qaeda, heightened fears of the terrorist–WMD linkage.[4]

When it came to proliferation and proliferation finance, al-Qaeda was not the only issue that at the time received attention in the press and among the arms control community. The global activities of the Pakistani scientist A.Q. Khan, who had for over twenty years managed an extensive procurement and financial network supported by shell and front companies aimed at acquiring nuclear technology (including from a raft of European suppliers), also began to feature prominently in counter-proliferation calculations. These came to recognise that the problem was far broader than that posed by terrorist groups such as al-Qaeda.

The clandestine financial network Khan created, starting in the 1980s, to support Pakistani procurement efforts was soon to be complemented by a financial network underpinning supply as Khan began to transfer nuclear technology to countries such as Iran, Libya and North Korea. Indeed, it was the interception by American agents of a German ship named *BBC China*, which was carrying nuclear technology headed for Tripoli, that heralded the downfall of Khan's proliferation empire.[5]

Even more troubling, knowledge of Saddam Hussein's WMD programmes and experience of the Gulf wars in 1991 and 2003 led many Western states to become extremely sensitive to the dangers of proliferation, the financing of WMD programmes, and especially the involvement of companies and financial institutions in Europe and elsewhere in supporting these projects.

Not surprisingly, the UN focus on terrorist organisations was ultimately complemented by its adoption of specific country proliferation measures.

UN Security Council Resolution (UNSCR) 1718 (2006) together with successor resolutions focusing on North Korea, and UNSCR 1737 (2006) together with successor resolutions (including 2231 (2015), which endorsed the Joint Comprehensive Plan of Action (JCPOA)) focusing on Iran, are core pillars in the United Nations' counter-proliferation architecture, including that for finance.[6]

These two countries have remained central to UN, FATF and national sanctions efforts in general and counter-proliferation financial efforts in particular. North Korea's first nuclear test in 2006 and Iran's nuclear weapons development efforts, together with their ballistic missile programmes and the financing involved therein, underpinned these concerns.

Indeed, although counter-proliferation finance is framed in general terms, it has emerged primarily as an anti-North Korean and anti-Iranian endeavour.

Counter-Proliferation Definitions and Recommendations

While the United Nations set broad objectives towards which member states needed to work, it did not attempt to provide detailed guidance as to how to achieve them. The emergence of counter-proliferation finance as a distinct financial strategy received its primary expression in the FATF Proliferation Financing 2008 Report.[7] The report relied upon a definition of proliferation finance which stated: 'Proliferation financing is providing financial services for the transfer and export of nuclear, chemical or biological weapons; their means of delivery and related materials. It involves, in particular, the financing of trade in proliferation sensitive goods, but could also include other financial support to individuals or entities engaged in proliferation.'[8] In 2010 this definition was refined as follows:

> Proliferation financing refers to the act of providing funds or financial services which are used, in whole or in part, for the manufacture, acquisition, possession, development, export, trans-shipment, brokering, transport, transfer, stockpiling or use of nuclear, chemical or biological weapons and their means of delivery and related materials (including both technologies and dual-use goods used for non-legitimate purposes), in contravention of national laws or, where applicable, international obligations.[9]

The 2008 Proliferation Report provided an example of how this proliferation financing would work, but essentially it referred to a common form of export financing indistinguishable from that used in countless daily legitimate transactions, albeit one that was being abused for proliferation purposes. So, an importer will organise the shipment of certain goods from an exporter. The importer may deal with the exporter directly or employ a cut-out such as a front company, broker or both. Again, payments will be settled with a manufacturer's bank either directly, by way of a front company or by employing a broker or both. The manufacturer may arrange the payment by way of a letter of credit or one of many other payment methods. The use of intermediaries is a means of disguising the end use, the goods themselves and the parties involved, but the use of intermediaries as the report explicitly admits is not in itself an indication of illegal intent as legitimate businesses rely on similar structures. Clearly the problem here that needs to be overcome is that the aim of the proliferator is to embed the transaction within a typical trade pattern so as to disguise the purpose of the deal. Not only will a financial institution have to distinguish between a legitimate and illegitimate trade activity, but the illegitimate one is being actively camouflaged to appear legitimate.[10]

The 2008 Proliferation Report differentiated proliferation financing from money laundering in that the sources of funds were often legal but the end-user was obscured. Thus it admitted that 'existing case studies do not enable us to identify any single financial pattern uniquely associated with proliferation financing,'[11] though such studies may help in identifying methodologies employed by proliferators.

It is important to recognise that FATF's approach to counter-proliferation financing did not require states specifically to undertake an assessment of their proliferation financing risk.[12] Moreover, the suggested approach was not risk-based but rules-based. The 2008 Proliferation Report thus did not recommend that organisations use a risk-based methodology but instead suggested that they should focus on the screening of designated entities already engaged with the firm. More ambitiously, however, organisations also need to engage in preventive action in relation to those designated entities as well. This arguably means that the requirements imposed upon compliance departments are rather more demanding than merely checking against lists of designated persons and entities. It involves something intellectually more proactive, including a knowledge of proliferation networks and processes. Of course, whether the relevant personnel involved in private sector compliance are sufficiently trained and able to undertake this kind of analysis is another matter entirely.

In order to assist firms in identifying proliferation risk in this very difficult context, FATF explicitly named various red flags. These include transactions involving individuals or entities in a jurisdiction of proliferation concern or in jurisdictions of diversionary concern. The use of these red flags would include staff checking against designated lists, a task that could be readily undertaken and potentially effective, assuming of course that the lists provided to the financial institutions are detailed and up to date. Some red flags involved more subjective or at least debatable judgements, such as whether the trade finance transaction involved shipment routes (if available) through jurisdictions with weak export-control laws or weak enforcement of export-control laws or involved individuals or companies (particularly trading companies) located in countries with weak export control laws and enforcement. Transactions involving shipment of goods incompatible with 'normal geographic trade patterns' are again something more difficult for all banks and exporters to determine. Requiring the financial institution or exporter to decide with certitude that the shipment of goods is incompatible with the technical level of the country to which the goods are being shipped is an equally challenging subjective judgement. Hard-pressed compliance departments certainly have some difficult decisions to make as they grapple with these counter-proliferation finance requirements.

The 2008 Proliferation Report provided the starting point for approaches by both the public and private sectors, and in 2012 FATF added proliferation financing to its list of standards and recommendations. Recommendation 7 stated:

Countries should implement targeted financial sanctions to comply with United Nations Security Council resolutions relating to the prevention, suppression and disruption of proliferation of weapons of mass destruction and its financing. These resolutions require countries to freeze without delay the funds or other assets of, and to ensure that no funds and other assets are made available, directly or indirectly, to or for the benefit of, any person or entity designated by, or under the authority of, the United Nations Security Council under Chapter VII of the Charter of the United Nations.[13]

This recommendation was supported by Recommendation 2, which required states to adopt national co-operation and co-ordination mechanisms to combat proliferation. In turn, these recommendations were buttressed by Immediate Outcome 11 and parts of Immediate Outcome 1, which sought to measure the effectiveness of the implementation of the relevant recommendations.[14]

Recommendation 7 thus links and applies to all current and successor Security Council resolutions applying targeted financial sanctions to the financing of proliferation of weapons of mass destruction. These included resolutions 1718 (2006), 1874 (2009), 2087 (2013), 2094 (2013), 2270 (2016), 2321 (2016) and 2356 (2017). Resolution 2231 (2015) endorsed the

Joint Comprehensive Plan of Action (JCPOA) and terminated all provisions of resolutions relating to Iran and proliferation financing, including 1737 (2006), 1747 (2007), 1803 (2008) and 1929 (2010).

As the guidance notes to Recommendation 7 stress, the focus is on preventive measures that are unique in respect of halting funds flows or the provision of other assets to proliferators or in support of proliferation. To this end states should designate targets to be subject to relevant proliferation-related sanctions, and dealings with these designated targets are prohibited and their assets are required to be frozen.

North Korea and Iran, as noted, unsurprisingly bore the brunt of all these counter-proliferation efforts. Resolution 2270 (2016) targeting Pyongyang focused on list-based sanctions but also included sanctions related to activities. It prohibited dealings with and required freezes in relation to entities linked to the government of North Korea and the Workers' Party of Korea that member states determined were associated with the country's nuclear or ballistic missile programmes or any other activities prohibited by UNSCR 1718 (2006) and its successor resolutions. It also required restrictions to be placed on financial dealings with North Korean banks based outside that country.[15]

The guidance note to Recommendation 7 sets out the criteria for designation as outlined in relevant resolutions (UNSCR 1718 (2006), 2087 (2013), 2094 (2013) and 2270 (2016)). In respect of North Korea these include, but are not limited to, persons or entities engaged in or providing support to the country's nuclear-related, other WMD-related and ballistic missile-related programmes; or assisting in the evasion of sanctions or in violating the provisions of UNSCR 1718 (2006) and 1874 (2009); or any entity of the North Korean government or the Workers' Party of Korea, or persons or entities acting on their behalf or associated with the country's nuclear or ballistic missile programmes or other activities prohibited by UNSCR 1718 (2006) and successor resolutions.

While some sanctions against Iran were lifted in 2016, those targeting Teheran's proliferation efforts remained very much in place. Relevant counter-proliferation finance criteria in relation to Iran were set out in UNSCR 2231 (2015) and included persons and entities engaged in or providing support for Iran's proliferation-sensitive nuclear activities contrary to that country's commitments in the JCPOA, or the development (including procurement) of nuclear weapon delivery systems; or those assisting designated persons or entities in evading or acting inconsistently with the

JCPOA or UNSCR 2231 (2015); or those acting on behalf of or at the direction of any prohibited person or entity. The re-imposition of American sanctions against Iran in 2018 did not materially impact upon counter proliferation sanctions which remained a constant.

Emboldened by its counter-proliferation actions in 2012, FATF returned to this theme the following year by fleshing out its guidance in a document entitled *The Implementation of Financial Provisions of United Nations Security Council Resolutions to Counter the Proliferation of Weapons of Mass Destruction*, in which it sought to divide UN counter-proliferation efforts into several baskets. These included the standard targeted financial sanctions and other financial provisions, and also the controversial activity-based prohibitions and the even more controversial vigilance measures which require a degree of pro-active awareness and action.[16]

Concerns expressed by the private sector in relation to these latter measures do not appear to have gained significant traction, though immediate outcomes in relation to counter-proliferation finance still did not formally require these entities to undertake risk assessments and mitigation steps such as those undertaken in the Anti-Money Laundering (AML) and Counter-Terrorist Financing (CFT) fields. Nor were there requirements for entities to undertake suspicious activity reports in relation to proliferation finance unless there were money-laundering issues involved as well. Nevertheless, these factors did not entirely square with the demand for vigilance, which implied a set of actions somewhat greater than list checking.

The Case of North Korea

The North Korean WMD programme together with its ballistic missile delivery project has always been central to counter-proliferation and counter proliferation finance efforts at the United Nations and at FATF and as expressed in Western (especially US) foreign and security policies.[17] In 2017 these sanctions efforts were intensified when the Security Council adopted four resolutions strengthening existing sanctions and initiated new measures such as petroleum sanctions against Pyongyang and a cap on crude oil. To the extent that sanctions efforts have succeeded or failed here—and by mid-2018 the view was that the regime in Pyongyang was increasingly under economic pressure—this was primarily due to steps taken against both imports and exports by North Korea, including trade in petroleum, crude oil and coal, especially with China, rather than any narrowly focused counter-proliferation financial actions, which appear to have achieved little to curb North Korea's nuclear ambitions.

In March 2018 the UN Panel of Experts, which had been established pursuant to UNSCR 1874 (2009), presented a report on its findings and recommendations in relation to sanctions against North Korea. Ten years after the FATF 2008 Proliferation Report, it was evident that there remained major problems in the implementation of sanctions capable of curtailing proliferation and related financing.[18] Indeed, as late as 2018 there existed significant deficiencies in the global implementation of financial sanctions, with the result that Pyongyang's evasive financial activities were neither being identified nor properly prevented. The report certainly painted a sorry picture, one characterised by the failure of sanctions and counter-proliferation finance.

Thus designated North Korean banks retained more than thirty overseas representatives who apparently remained free to move across Asia and the Middle East and who facilitated transactions, controlled bank accounts and dealt in bulk cash transfers. The Panel of Experts criticised member states who were, they claimed, not sufficiently scrutinising the activities of North Korean bank representatives nor, when they were identified, acting to expel them. Nor were branches, subsidiaries or representative offices being closed or relevant assets frozen. There were significant deficiencies at the on-boarding stage of new clients while accounts of existing clients were not being properly monitored. North Korean accounts continued to be opened in the names of foreign nationals, family members and front companies with a North Korean national authorised to have signing rights and ultimate control of the accounts. Bulk cash withdrawals continued to take place, even though they had long been recognised as an important method of financing North Korean proliferation transactions. North Korean diplomatic property was in several instances used for commercial purposes, including those specifically prohibited by sanctions.

Rather than serving as gatekeepers, various corporate service providers acted as facilitators and helped North Korean financial institutions to establish offshore front companies which employed non-North Korean nationals to abet global financial transfers. Joint ventures with foreign companies helped generate foreign currency. North Korean companies such as MKP and Glocom were, according to the UN study, linked through a complex chain of front companies to the Reconnaissance General Bureau, a North Korean intelligence agency.

MKP, for example, had developed a network in several continents and was involved in multiple industry sectors including mining, coal trading, security, construction and transport, with more than fifteen affiliated companies in four jurisdictions. The way in which MKP established an office in South Africa was characteristic of its approach, including the use of embassy officials to initiate business introductions and the use of falsified documents to skirt local due diligence requirements.

Glocom, in turn, was tasked by the Reconnaissance General Bureau with marketing military communications technology. The company controlled more than ten accounts in several states, which allowed Glocom to shift funds readily across borders. Glocom's financial transfers within its network for the purpose of payments to suppliers were typical of how North Korea appeared to evade counter-proliferation finance strategies aimed at stopping its business. In the first stage a North Korean national would make a bulk cash deposit into a Glocom account held in Pyongyang. The deposit would then be reconciled by ledger with an account at a bank in China. On the same or the next day, funds would be transferred to recipients in Singapore or Malaysia for the same amount by way of a Hong Kong-based front company. The receiving entities in Singapore and Malaysia would see the incoming payment as deriving from the Hong Kong front company rather than the holder of accounts with the Chinese bank. Similarly, the correspondent banks processing the transaction, including those based in New York clearing any dollar transactions, would not be able to see either the originators of the funds or the beneficiaries.

Among the suggestions for improving financial sanctions, the Panel of Experts recommended that member states implement a risk-based approach for identifying sanctions violations in their 'know your customer' and compliance programmes, including more robust on-boarding, transaction monitoring and account review practices. They suggested that member states should provide more detailed information to financial institutions and that more attention be paid by corporate service providers to implementing due diligence aimed at revealing the beneficial ownership of front companies. The panel called on member states to put in place an appropriate legislative framework for combating proliferation financing and, significantly, demanded the establishment of robust information-sharing mechanisms among financial intelligence units, customs and strategic trade control authorities, border control, and security and intelligence agencies, and with the private sector.[19]

In addition to a focus on financial sanctions as a means of constraining North Korean proliferation programmes, global efforts also placed much emphasis on containing that country's illicit shipping. Concerns about these

North Korean tactics led the US Treasury in 2018 to announce its 'largest North Korea sanctions tranche to date', covering 56 firms and vessels linked to North Korea. All these entities and vessels were subjected to an asset freeze and prohibited from dealing with US persons.[20] Coal makes up 40 per cent of North Korea's foreign income.

Pyongyang has certainly employed illicit shipping to import petroleum and export-sanctioned commodities. To these ends it relies upon the use of fraudulent documentation, deceptive navigation patterns, signals manipulation and other methods to disguise the origin of the relevant commodity. Shipping was used, as one diplomat put it, to 'launder the coal … it's the same tactic criminals use to launder ill-gotten cash'.[21]

This multi-million-dollar business involves a global network of ship charterers, brokers and global commodity trading companies, some of which may not be fully aware that they are complicit in sanctions busting and the support of proliferation activity. Again, according to the Panel of Experts, the usual suspects are to be held accountable and the private sector is asked to step up its game. The report states: 'The profit margins involved, coupled with the offshore nature of much of the affected oil, maritime and finance sectors, necessitates far greater private sector due diligence, information-sharing and self-policing.'[22]

Given the requirements made of the private sector to help clamp down on these illicit activities, a debate will need to take place as to whether the efforts involved in this significant detective work match the scale of the ultimate threat. The Panel of Experts, for example, claims that between January and September 2017 these illicit activities generated nearly $200 million. In the greater scheme of things this is not an overly large sum, and it is minuscule when set against the needs of the North Korean economy. The main body of water which certainly requires monitoring is not the open seas but the Yalu River, across which much of North Korea's key trade with China takes place.

In the final analysis the Panel of Experts report noted:

> The international community has responded to the unprecedented nuclear and ballistic missile activities of the Democratic People's Republic of Korea in 2017 by introducing an array of new measures that seek to change the country's behaviour and direction in 2018. However, this expansion of the regime is yet to be matched by the requisite political will, international coordination, prioritization and resource allocation necessary to drive effective implementation.[23]

Indeed, while arguably by mid-2018 sanctions helped bring North Korea to the negotiating table with President Trump, they had not halted or effectively impacted on that country's WMD programme which according to some analyses already boasted by then both thermonuclear devices and rudimentary intercontinental delivery capabilities.

In the North Korean instance, at least, sanctions proved more useful as very blunt instruments rather than as refined tools able to restrict specific capabilities.

The 2018 FATF Counter-Proliferation Finance Report

Given the challenges posed by North Korean and Iranian proliferation activities, proliferation financing continued to receive significant attention by policy-makers, who felt that the finance community should be taking a more active role. In 2018 the FATF Proliferation and Financing Report of 2008 was updated with new guidance on counter-proliferation finance.[24] The new report fleshed out and updated some of the themes originally addressed in 2008 but in many respects it represented a continuity with the earlier report rather than bringing to the table any revolutionary new approaches.

The 2018 CPF Report provides non-binding guidance in four main areas: implementing UNSC targeted financial sanctions; implementing other measures included in Security Council resolutions; improving information sharing and inter-agency co-operation; and supporting national authorities with compliance and monitoring. It also expands and updates various red flags that can help institutions in both the public and private sectors in identifying proliferation activities.

Targeted Financial Sanctions

The 2018 CPF Report begins by including guidance on the implementation of relevant Security Council resolutions. Specifically, states are required by Recommendation 7 to comply with relevant resolutions and implement sanctions relating to WMD proliferation. Sanctions are applicable to those persons and entities designated by the Security Council or by its relevant committees. Designation criteria are described in the guidance notes in Recommendation 7, and states are required to freeze funds, other financial assets and economic resources in other territories or under their jurisdiction. There are two country-specific regimes: North Korea and Iran. In the case of North Korea, the designating authority is the Security Council and its 1718 Committee, and in the case of Iran it is the Security Council only as the relevant committee has been terminated following UNSCR 2231 (2015). Entities in both the public and private sectors will need constantly to screen designated persons and entities. As the 2018 CPF Report admits: 'However, the screening would not be sufficient on its own, as targeted financial sanctions are also applicable to persons/entities acting on behalf of or at the direction of designated persons/entities. This adds additional complexities for public and private sector entities in identifying and detecting the persons, entities, and transactions related to proliferation financing.'[25] How private sector entities will secure such information is unclear, but the 2018 CPF Report notes that intelligence agencies, law enforcement agencies and financial intelligence units will need to share their information with the private sector. This undoubtedly will be key to the proper identification of proliferators, a process that will also require state-to-state co-operation.

Central to the development of national counter-proliferation strategies is the implementation of co-ordinated measures to identify and prevent the evasion of sanctions. In support of this, FATF provides a non-exhaustive list of contextual factors that may help indicate efforts to evade sanctions. The first of these concerns the volume of international services, which means attention needs to be paid to financial centres where financial service volume can be used to hide proliferation finance activity. Secondly, note must be taken of the level of AML and CFT compliance as proliferation financing is attracted to weak legal and regulatory frameworks. Thirdly, there is the underlying proliferation risk, which involves identifying geographic vulnerabilities, the presence of persons from countries subject to sanctions, industries producing dual-use proliferation-sensitive or military goods, and international shipping and trade services. Also relevant is the strength of export controls, and customs and border controls, and whether these are effectively implemented so as to mitigate proliferation risk. Unfortunately, some of these identifying markers are so general as to limit their potential effectiveness in narrowing down the field. For example, the requirement to focus on financial centres is not commercially practical given the amount of trade that by definition flows though such hubs. A major bank will have countless transactions daily involving places such as Dubai and Singapore, not all of which can be subjected to heightened due diligence.

These measures will of course need to be viewed alongside other information provided to the private sector so as to help in the identification process. Information of this kind is the key to effective private sector counter-proliferation finance strategy. This is recognised by FATF, which suggests that the competent authorities provide additional relevant information to supervised institutions, including names of specific entities and persons who may be tied to proliferation networks. Furthermore, authorities need to provide institutions with information regarding end-users of concern in respect of technology, equipment, goods and material prohibited under specific Security Council resolutions relating to the spread of WMD and also lists provided by relevant national export control authorities. Other information that should be provided includes that dealing with the diversion of items prohibited under country-specific resolutions and lists as well as the characteristics of persons who have been granted or denied export licences.[26] Armed with this type of information supplied by state authorities, private sector institutions will be better able to act effectively.

FATF reinforced the North Korean focus when it emphasised in its report that in relation to Pyongyang countries should be aware of factors such as withdrawals or deposits of bulk cash, the opening of bank accounts by North Korean diplomats, the clearing of funds, and the provision of export credits or guarantees to persons or entities transacting with North Korea. Significantly, attention needed to be placed on North Korean maritime activities, which, as sanctions against Pyongyang tightened, especially along its northern border with China, were increasingly used to circumvent these growing restrictions. This included the provision of insurance or reinsurance services to vessels controlled by North Korea which might be involved in transporting items prohibited by the Security Council, the direct or indirect supply or transfer to North Korea of vessels, and the leasing or provision of crew services to the country.[27]

Other Measures Included in UNSCRs

The 2018 CPF Report recognises that Security Council resolutions can reach parts of the proliferation finance process not addressed by FATF and not assessed in FATF mutual evaluation exercises. It recognises that these resolutions go beyond Recommendation 7 but also impact on counter-proliferation financing. Examples given are the mandatory UNSCR obligations such as activity-based financial provisions and economic or sectoral sanctions.

Activity-based financial prohibitions require states to forbid certain financial activities connected with proliferation, with a specific focus on North Korea and Iran. We have already touched upon some of these, but they include, in relation to Iran, the prevention without Security Council approval of any financial assistance, investment or other services. Also prohibited is the acquisition of an interest by Iran in any commercial activity in another country in relation to items set out in the Missile Technology Control Regime and the Nuclear Suppliers Group, or the provision of financial resources to support the transfer, manufacture or maintenance of certain conventional arms as defined on the United Nations Register of Conventional Arms. In relation to North Korea, activity-based financial prohibitions include those directly touching on proliferation, such as the ban on the provision of financial services that could contribute to WMD proliferation, or on financial support for the procurement of missiles and related technology, together with more general sanctions already alluded to, such as the prohibition of the opening and operation of bank branches and subsidiaries, the transfer and use of bulk cash, and the provision of financial support for trade.[28]

In addition to activity-based financial prohibitions, the 2018 CPF Report refers to Security Council resolutions requiring states to exercise 'vigilance' and prevent North Korean procurement of technologies underpinning proliferation or financial services in support of proliferation. Nor are member states to enter into new loans or other forms of financial commitment with North Korea.[29] Finally, both the activity-based financial prohibition approach and the vigilance measures are accompanied by Security Council economic and sectoral sanctions which prohibit the trading of various goods and material that could contribute to the proliferation of delivery systems, including aircraft, gold, titanium, petroleum products, iron, iron ore and coal.[30]

The 2018 CPF Report sets out various general principles that states should follow, including how they should deal with supervised institutions in the private sector. There is clear continuity between these approaches and AML, CFT and general sanctions approaches, except that here the goal is to counter proliferation. Thus states and private institutions working in the context of counter-proliferation need to identify high-risk customers and transactions, enhance scrutiny of high risk customers and transactions, and to set in place procedures for follow-up actions. Specifically:

> Countries may encourage supervised institutions to consider proliferation financing typologies when reviewing transactional and customer information collected through their existing AML/CFT obligations and customer due diligence programmes. The information collected should allow supervised institutions to identify transactions, accounts (including correspondent accounts), or relationships (such as accounts of DPRK diplomats, joint ventures or jointly owned banking operations or facilities) with Iranian and DPRK banks and entities.[31]

The Requirement for Counter-Proliferation Finance Co-operation

Because of the serious challenges involved in identifying relevant activities, co-operation is crucial between businesses in the private sector and those state authorities holding relevant information in relation to the counter-proliferation process. The 2018 CPF Report admits that lack of co-ordination is a common deficiency among FATF members. Some countries have neither the legal nor the institutional framework to implement FATF guidelines in dealing with counter-proliferation financing while others have not established agencies or committees to supervise enforcement. Nor in some instances have any measures been undertaken to co-ordinate counter-proliferation efforts. In these respects it is indisputable that counter-proliferation efforts have fallen behind other areas dedicated to the financial aspects of security.

The 2018 CPF Report helps to identify those national agencies that should be given a key role in managing the counter-proliferation finance process. These include policy departments, financial supervisors, competent authorities, export control and customs agencies, intelligence services, financial intelligence units, law enforcement and prosecution agencies, and trade promotion and investment agencies. Working together, these agencies and departments need to develop a co-ordinated approach for communicating with entities in the private sector. Central to this co-ordination should be the provision of updated lists of designated persons, entities and vessels upon which the private sector can act. As the 2018 CPF Report notes:

> Information on suspected proliferation financing activities gathered by counter proliferation authorities, as well as [financial intelligence units'] feedback on [suspicious transaction reports] which eventually point to proliferation financing activities should also be shared with supervised institutions as appropriate and subject to countries' existing legal frameworks. This will help private sectors detect proliferation financing sanctions evasion early.[32]

The United Kingdom provides a useful example of counter proliferation finance cooperation at both the intra-governmental and public-private levels. The commitment of the country to counter proliferation was set out in the 2015 Strategic Defence Review while the National Counter Proliferation Strategy 2020 emphasised that counter proliferation finance was a key element of that strategy. The Counter Proliferation and Arms Control Centre was established in 2016 and co-ordinates a range of ministries working on counter proliferation including counter proliferation finance. There are various other relevant intra-governmental co-ordinating mechanisms which also work towards proposing proliferation finance designations. So for example in 2017 the United Kingdom proposed at the European Union level various DPRK designations and measures beyond those adopted by the UN.[33]

The Office of Financial Sanctions Implementation (OFSI) which we discussed in the previous chapter plays an important role in enforcing breaches of proliferation related sanctions. It works with the private sector in relation to such sanctions and also co-ordinates with the National Crime Agency when sanctions are breached. In 2016–17 109 cases of sanctions breaches were reported to OFSI with a potential value of £227 million. The Joint Money Laundering Intelligence Task Force which we also previously reviewed serves an important co-ordinating function between the public and private sectors in relation to proliferation finance issues. Intelligence is shared and feedback is received. FATF in its 2018 review of the United Kingdom's proliferation finance policies and structures gave the system high marks but noted that an understanding of the risks of proliferation finance varied across the financial industry and was found most lacking in smaller organisations.[34]

Supervision

In addition to the lack of co-ordination that is endemic to counter-proliferation financing activities, another key deficiency that has been identified is unfocused supervision. The 2018 CPF Report sets out the basis for an effective supervisory model, which includes recommendations for the control and monitoring process, suggested remedial actions and sanctions, and the need for competent authorities to promote among supervised institutions an understanding of their obligations in the counter-proliferation field.[35]

Counter-Proliferation Finance Red Flags for Public and Private Sectors

The 2008 Proliferation Report had highlighted various red flags that serve to enable both public and private sector entities to identify proliferation finance behaviour. By 2018 proliferation financing techniques had evolved. Some of these were identified in reports by the UN Panel of Experts to various Security Council committees as well as by state authorities and academic institutions. However, the 2018 CPF Report admitted that the red flags identified were not necessarily in themselves determinative of proliferation financing and that there was overlap between proliferation-financing activities and terrorist-financing and also money-laundering activities, especially trade-based money laundering. In the report there emerged twenty of these red flags, consisting of ones already covered in the 2008 Proliferation Report and some new ones. The problem of differentiating illicit from entirely legal activity remained, however, because while some red flags were clear and readily identifiable—for example the involvement of items controlled under WMD export control regimes or national control regimes or evidence that documents or other representations are fake or fraudulent—others were of a more general nature and in themselves (without further relevant information) indistinguishable from legitimate trade. Examples of the latter include:

- Transactions involving persons or entities in foreign countries of proliferation concern;
- Transactions involving persons or entities in foreign countries of diversion concern;
- A freight forwarding firm is listed as the product's final destination;
- Circuitous routes of shipment (if available) or circuitous routes of financial transaction;
- Trade finance transactions involving shipment routes (if available) through countries with weak export-control laws or weak enforcement of export-control laws;
- Transactions involving persons or companies (particularly trading companies) located in countries with weak export-control laws or weak enforcement of export-control laws;
- Transactions involving shipment of goods inconsistent with normal geographic trade patterns;
- Involvement of a university in a country of proliferation concern;
- Use of personal accounts to purchase industrial items.[36]

Some of these red flags will no doubt be challenging for private sector entities. Transactions involving persons or entities in countries of proliferation concern are way too general for compliance departments to focus on unless they possess specific names. For example, India and China—two countries of proliferation concern—are involved in countless daily global transactions, making it impossible to monitor each such transaction with detailed scrutiny. Presumably also relevant here would be Israel, Egypt and Pakistan, states whose trading volumes are not themselves insignificant. Unless banks or companies are provided with specific names of individuals or entities, the task is far too massive. And even with that information, due diligence processes at the client on-boarding and account or transaction-monitoring phases are expensive, time-consuming and far from infallible when determined proliferators are involved.

Again, the reference to red flags linking to 'involvement of a university in a country of proliferation concern' is extremely general and raises clear definitional issues. Why would fraudulent representations be an indication of proliferation concern as opposed to other illicit activities? How would an entity based in London know that a small intermediary company is carrying out business that is incompatible with their normal business? Of course, many of these issues could ultimately at least be determined with some degree of certitude, but the level of due diligence required would in many cases be excessive, involving amounts of money and time that would render the transaction uncommercial.

An analysis of some of these red flags leads to the conclusion that while undoubtedly well intentioned, they do not always appear grounded in commercial realities pertaining to global trade and global finance.

Given these problems, it is not surprising that some analysts have been quite critical of the FATF's counter-proliferation guidance, questioning whether it is fit for purpose when it comes to countering proliferation finance. Aaron Arnold has written that the guidance provided by the 2018 CPF Report does 'little to improve global implementation of rules and regulations specifically focused on proliferation financing—implementation that today is half-hearted' and that 'the new FATF report contributes little to international efforts against proliferation financing'.[37] Arnold admits the reality of the situation when he goes on to conclude, correctly, that 'addressing proliferation financing is very complex, and in some cases it may not be possible'—certainly when the requirements go beyond checking lists (presumably employing relevant software) to involve preventive measures.[38]

How it can be possible for compliance professionals functioning within a highly pressured, time-urgent and very competitive context to effectively and readily distinguish between proliferation activities and legitimate trade financing is something that needs more attention.

In his discussion of North Korean proliferation techniques, Arnold notes that none of them are particularly innovative and that there is a continuity between these techniques and those of sanctions evaders, drug traffickers and others. He suggests that the problem may be less one of specific counter-proliferation finance implementation problems and more reflective of broader systemic weaknesses, such as the existence of exploitable offshore jurisdictions. Here he does not sound too optimistic a note about remediation efforts and, unfortunately, he is probably right about that as well.

These findings are in several respects supported by the detailed work conducted by Tom Keatinge, Emil Dall and Andrea Berger at the Royal United Services Institute (RUSI) in 2016. This research does not paint a picture of a public sector properly organised or a private sector fully convinced of the merits of counter-proliferation financing. From the state sector there is mixed messaging expressed in indicators, typologies and red flags which vary in degrees of detail and usefulness. There are problems of vagueness, contradiction and overlap with other areas. Some red flags cannot be actioned by the private sector not only because of lack of access to the required intelligence but 'because doing so would require enormous resources and technical expertise'[39]—in many senses this is the crux of the matter. They report that there are differences in views among governments as to the feasibility of using private sector institutions as a means of defence against proliferation and the priority that should be accorded to proliferation finance over other financial security risks. While there is an admission that 'proliferation does not end where the listing stops', checking against lists is already quite challenging.[40] And then, of course, should the challenges prove too great, there is always the possibility of de-risking by institutions, which in turn has its own financial security dangers.

Another fundamental problem is identified by the authors of the RUSI report when they write that financial institutions 'on the whole do not understand the contemporary realities of the threat they are facing. Few financial crime representatives interviewed for this report understood what proliferation involves in practice.'[41] This is unsurprising as it is not an area of expertise for which compliance professionals are trained. Many banks and other institutions choose to shy away from a goods and an activity focus, and prefer to employ third-party external software providers to help them focus on entities in the context of general sanctions—especially American sanctions. As the RUSI report goes on to argue: 'To effectively conduct screening on an activity basis, [financial institutions] must consider the tactics used by proliferators to conceal their illicit aims, broader proliferation trends and the way in which these dynamics might manifest themselves within global financial flows.'[42] This, to put it mildly, will be an extremely difficult task for many compliance departments or front offices of banks, let alone smaller import-export entities.

Keatinge and his team recognise these problems and seek to provide better and more detailed guidance for financial institutions to cope with proliferation threats to the financial system and specifically their own institutions.

They cogently argue that while proliferation finance may share characteristics with other forms of financial crime, it has its own distinct characteristics. They go beyond a focus on FATF requirements and call on financial institutions to conduct internal risk assessments of their exposure to proliferation finance. Beyond consultation with red flags, case studies and knowledge of export-control lists, banks and other financial institutions should monitor the end use to which these products are put. More controversially they argue: 'Financial institutions should take measures to move beyond focusing merely on the entities and individuals on sanctions lists. This includes conducting network analysis in order to determine whether any clients are connected to or doing business with designated entities.'[43]

While there could be undoubted benefits to such investigations, it will obviously also be extremely challenging for banks, financial institutions, export companies and other institutions to involve themselves in this kind of 'network analysis', given the way they are structured, staffed and function within the relentless commercial hothouse of banking and finance. Where the RUSI proposals are undoubtedly at their strongest is when they call for better information sharing between the public and private sectors, as much information goes wasted and unexploited. Better, more detailed, relevant and actionable lists of persons and entities will go a long way towards helping financial institutions become effective in the counter-proliferation field.[44]

Scepticism about a Counter-Proliferation Focus

Top Cape Technology was a rogue company exploiting vulnerabilities within the South African export control system. But under the previous apartheid regime, clandestine proliferation programmes were government-initiated and -managed. In 1989 South Africa terminated its nuclear weapons production efforts and dismantled all six of its nuclear weapons while plans to produce a seventh were also ended. Then in 1991 the country went on to sign the Nuclear Non-Proliferation Treaty. But during the previous two decades, as South Africa faced growing international isolation and fears that the Soviet Union would become increasingly involved in the southern African region, the country had engaged in the development and financing of a major nuclear weapons build-up. This sophisticated and extensive production and financing programme had domestic and international dimensions and took place in the forbidding context of UNSCR 418 (1977), which had imposed a mandatory arms embargo on the apartheid regime and had called on all states to refrain from 'any co-operation with South Africa in the manufacture and development of nuclear weapons'.[45]

So when in 2010 and 2011 FATF held a series of meetings to consider issuing an updated set of recommendations in 2012, including a new proliferation finance recommendation (which became Recommendation 7), the South African attendees clearly had experience upon which to base their contribution to the proceedings. Arguably, they had more direct knowledge of this field than any other participants in the various evaluation rounds that took place. The South Africans, to put it mildly, were very sceptical about the sagacity of investing effort in the counter-proliferation space and in dealing with proliferation finance on an activity basis. In prescient language, the Banking Association of South Africa stated that 'the financial sector is unable to provide any support to governments' anti-proliferation efforts due to the difficulties associated with "dual use" items. We again recommend that the FATF not engage itself or its member in this impossible effort, which is best left to trained and experienced customs officials.'[46]

The South Africans were not alone in their scepticism. Other countries voiced concerns about being able to distinguish between proliferation finance techniques and general trade activities. Germany—a country with several companies which had been found complicit in Saddam Hussein's nuclear and other WMD programmes—strongly articulated its opposition to onerous counter-proliferation finance obligations in addition to the already demanding obligations dealing with other threats to financial security. The European Banking Industry Committee also stated unequivocally: 'with regard to the extension of FATF requirements to proliferation financing, European banks regard an entity-based approach (checking lists of natural and legal persons) as the only feasible and cost effective option when it comes to combating proliferation finance ... General vigilance is impossible to implement for financial institutions.'[47]

Within the field of finance and security, the distance between the demands for private sector action and commercial reality is at its widest in the counter-proliferation finance space. The proactive requirements of financial and other institutions are difficult to implement, involve massive efforts and yield (as far as we can tell) modest results. The problems of differentiating legitimate from illegitimate trade finance activities and effectively screening dual-use goods are beyond the best intentions of many private sector entities. Private sector financial institutions would be best advised to focus their main efforts on ensuring they are not breaching sanctions and embargoes—especially US sanctions and

embargoes—and busy themselves with checking against the various lists provided by national authorities. Countering proliferation finance would be best
advanced by grounding responses within broader sanctions policy.

Counter-proliferation finance as a technique has proven less than effective
in supporting counter-proliferation policy. It is an unfortunate reality that as
a weapon of foreign and security policy, counter-proliferation finance may be
in many respects, at least in its present form, a bridge too far.

CONCLUSION

Major intelligence failures are usually caused by failures of analysis, not failures of collection.

Richards J. Heuer Jr

Costs and Expectations

In his classic study *Psychology of Intelligence Analysis* Richards J. Heuer Jr argues that when analytical judgements turn out to be faulty, it is most often the case that this was not because the information was wrong but because analysts held faulty assumptions that went unchallenged.[1] Are there central assumptions underpinning the financial strategy aimed at combating criminals, terrorists and proliferators, based upon a coalition of public sector bodies and institutions functioning in the private sector, that need to be questioned, and if so why?

Supporters of financial warfare have long emphasised the benefits of using tools of finance in countering criminal laundering, terrorist financing, proliferation financing and other threats to security in general and the integrity of the financial system in particular. The former US Treasury general counsel David Aufhauser, who in the period following the 9/11 terrorist attacks had significant anti-money laundering (AML) and counter-terrorist financing (CFT) responsibilities, claimed in 2003: 'Stopping the flow of money to terrorists may also be one of the very best ways we have to stopping terror altogether.' While Aufhauser did slightly qualify this comment with the statement 'We believe that if you stop the money, you go a long way to stop the killing',[2] his view is generally regarded as an expression of the centrality and importance of finance as a counter-terrorist weapon.

One first step in analysing the merits of these kinds of assumptions must be to assess in general terms the costs of what is being undertaken and the expectations surrounding those efforts.

This book has argued that at the core of the financial war on criminal money laundering and terrorist financing has been a government partnership with the private sector, especially financial institutions, designated non-financial institutions and various gatekeepers, such as lawyers and accountants, conscripted into the fight. Faced with the significant financial, reputational and other costs of not aligning themselves with government policy, the private sector has little choice but to comply, even when the gains of the campaign against criminals, terrorists and proliferators appear so paltry and even where the national authorities with whom compliance is demanded are foreign.

But have the gains indeed been marginal? The starting point must surely be the balance between the costs and results of the campaign. The British Bankers' Association has estimated the costs of general compliance for the United Kingdom alone at about £5 billion a year (other estimates are even higher), a not insignificant sum. But, as has been demonstrated in this study, to the extent they are measurable, the results of this war have been meagre or questionable, to say the least.

David Fein, a group general counsel at Standard Chartered Bank, has for example written: 'When it comes to money laundering, the criminals are winning. More than $1tn of illicit financing moves through the financial system each year. Only 0.2 per cent of those proceeds are seized by authorities, making the financial crime industry the largest and most profitable in the world.'[3] This sweeping assertion certainly seems to have some support if we look at a range of indices from conviction rates to asset freezes (see Chapter 5).

So much for the financial war against criminals. The financial campaign against terrorism also appears to be of limited value as the costs (as we noted in Chapter 2) are increasingly large and the terrorists often seem to avoid the formal banking system altogether. Banks continue to invest enormous sums in relevant capabilities, but examples of where finance as a weapon has directly halted or deterred terrorist attacks are difficult to determine and are certainly not extensive.

Where terrorists have gained control of states, the historical record (see Chapter 2) tends to suggest that it has been traditional land and air operations conducted by military forces that rolled up the terrorist state, rather than peripheral financial efforts carried out by a public–private financial partnership.

Successful efforts aimed at using finance to counter proliferation have also been limited, not least because such a campaign is often beyond the capabilities of much of the private sector. Counter-proliferation has had far more success in the field of direct military actions (for example, Israel's attack on Syrian nuclear development capabilities, and Israeli and Western attacks on Iraq's weapons of mass destruction arsenal and programmes) than any effort by public and private institutions to use finance to inflict damage on proliferation networks.

In the established and extensive literature that debates the effectiveness of sanctions, it is not surprising that assessments of the impact of offensive financial weaponry, such as financial and sectorial sanctions as well as embargoes, are not without controversy. At the time of writing, despite the imposition of a wide range of sanctions and embargoes, Russia still maintains its position in Crimea and eastern Ukraine, North Korea still retains its nuclear weaponry, and Iran at the very least its own missile force, a nascent if frozen nuclear programme and an aggressive regional policy. While (as noted in Chapter 6) sanctions and embargoes impose undoubted technical and financial costs on target countries, these do not readily appear to advance the primary objectives, be they territorial withdrawal, a decrease in foreign intervention or disarmament.[4]

In respect of disarmament, part of the problem is that the decision to acquire nuclear weapons is not at its core a financial one, but rather a reflection of political, strategic and, in some circumstances, existential considerations. In the 1970s, the Prime Minister of Pakistan, Zulfikar Ali Bhutto, famously said: 'We will eat grass, even go hungry, but we will have our own [nuclear weapons].'[5] Such decision-making, while not impervious to financial pressures, will be highly resistant to them.

The obvious disjuncture between costs (money and effort) and results is starting to give rise to a vigorous debate as to the efficacy of the central elements of finance warfare.

Professor Neumann and His Critics

In an article in *Foreign Affairs* in July/August 2017 entitled 'Don't Follow the Money', Professor Peter Neumann powerfully argued that 'the war on terrorism financing has failed'. Despite all the efforts involved, there is no evidence that the financial war has ever thwarted a terrorist attack. Although governments always turn to the financial system to crack down on terrorism, this usually fails because terrorists often don't have bank accounts in their own names and as a result banks are unable to identify suspicious transactions. Secondly, large amounts of terrorist finance in countries such as Afghanistan, Iraq, Somalia, Syria and Yemen never enter the global financial system in the first place. Neumann goes on to question the very assumption of there being a common set of financial methods that all terrorists employ. He points out: 'After spending over 15 years and billions of dollars on a strategy that has had little discernible impact, it's past time for a new approach.' Rather, states should, in his view, move away from an excessive delegation of responsibilities to the financial and banking sector and instead 'integrate their efforts to restrict terrorist financing into their wider counter terrorism strategies'.[6]

The thrust of this book has been generally supportive of Neumann's ideas and has adopted a somewhat sceptical attitude to the demands that national authorities increasingly foist upon the private sector in the fight against a range of targets (and not only terrorists). Critics of Neumann's view, however, quickly seized upon his essentially modest suggestion for improvement—better integration of efforts—and pointed out that such integration does in fact already exist. Matthew Levitt and Katherine Bauer, both former US Treasury officials who have been involved in the use of financial weaponry in fighting terror, called Neumann's approach 'bad advice' and said that while there was obvious room for improvement, Neumann's suggestions would effectively involve throwing the 'baby out with the bath water'. They go on to argue that financial tools in themselves are not meant to solve the problem but form part of a broader strategy, and that there have indeed been successes, including the gathering of financial intelligence that can be used by other arms of the state.[7]

Other Neumann critics joined together to write a *Foreign Affairs* article entitled 'Can Bankers Fight Terrorism?' and levelled similar critiques, including the very valid and central argument that the effect of the financial war on terrorism has rendered terrorist financing far more difficult than it otherwise would have been without all these concerted efforts. According to Professor Jodi Vittori's rather lukewarm praise, while the war on terrorist financing has not lived up to its promises, clearly it 'has done some good'.[8]

Arguably, it is in measuring the good that Vittori refers to against the costs incurred that the problem is to be found. Neumann's critics are correct in pointing out that some of his proposed solutions are already in operation and that there have been successes in the field, but his riposte to these criticisms has undoubted merit when he emphasises the opportunity costs that this financial focus must have incurred: 'Is the enormous bureaucracy that has been created in the name of countering terrorist financing, a bureaucracy that has imposed billions of dollars of costs on governments and the private sector, justified by its result? What else could have been done with all that time, effort and money?'[9] One cannot dispute the fact that financial intelligence played a role in helping identify Islamic State targets, but was the campaign to defeat the group really speeded up by financial targeting? Have al-Qaeda activities really been constrained by a focus on its financial activities and, as Neumann asks, how do we ultimately measure that effectiveness?

It is of course interesting that when *Foreign Affairs* chooses to publish an article of criticism under the title 'Can Bankers Fight Terrorism?' it does not include the views of any banker (or associated finance professional) but focuses mainly on the perspectives of those with links to the state. This is not to suggest that the views of these former state officials and academics are not uniformly persuasive and meritorious, but it is like analysing the effectiveness of a military strategy without consulting the views of military professionals.

Of course, given the febrile atmosphere surrounding terrorism and banking and the banking of terrorists, as well as criminal money laundering and weapons of mass destruction proliferation, the journal was perhaps hard pressed to find individuals in the finance sector who would be prepared to provide a critique of the increasingly heavy demands that continue to be placed on the sector and also their views of its efficacy. No one wants to take up a position of complaint about these escalating demands—and then something goes terribly wrong in the counter-terrorist, counter-proliferation or anti-money laundering fields. The sensitivity of financial institutions to public relations and reputational risk has risen over the years as banking activities have fallen more and more under the public spotlight.

Over the past few years there have emerged forums where government experts and the financial community can exchange views and perspectives. In the United Kingdom there is, for example, the Joint Money Laundering Intelligence Taskforce. In the United States there is an annual Public-Private Analytic Exchange Program convened by the US Department of Homeland Security and the US Office of the Director of National Intelligence which facilitates a public–private exchange of views. The US government also takes care to canvas the views of groups such as the Association of Certified Anti-Money Laundering Specialists and the Association of Certified Financial Crime Specialists while the Financial Intelligence and Information Sharing Working Group serves to share information between the private and public sectors. Views are canvassed and taken into account, but the steady stream of compliance demands remains relentless. And no one should be under any illusion that these are actually partnerships of equals—nor admittedly can they ever be.

Challenges Facing the Private Sector

For many financial institutions, AML and CFT and related compliance is thus viewed as essentially a tax that simply has to be borne though the inefficiencies and redundancies are plain for all to see, and many question the overall effectiveness of much that is undertaken. More galling is the sometimes hypocritical approach of government authorities, who are quick to demand but who sometimes adopt positions that seem to undercut the very demands made on the private sector. For example, the historically relaxed attitude of the British government to less-than-savoury overseas investments in UK markets as well as its tolerance of offshore banking and opaque structures goes hand in hand at the same time with a demand for increasing AML and CFT vigilance by the private sector—an incongruity that is only slowly being overcome.

Not that the private sector is without its faults in this regard: in many instances the sector has been too quick to embrace lucrative though shadily sourced fees. But sometimes, when the state—which after all creates the environment in which the private sector functions—comes under pressure, it is not beyond turning on the latter to shift some of the blame.

Thus when one of the most prestigious and highly respected law firms in the City of London declined to appear before the House of Commons Foreign Affairs Committee to talk about its work for a Russian client company which had recently floated on the London Stock Exchange (doing so would be a clear and fundamental breach of its client confidentiality requirements fundamental to its practice) one member of the committee claimed that although there was 'no obvious evidence' of impropriety, it 'must leave others to judge whether their work "at the forefront of financial, corporate and commercial developments in Russia" has left them so entwined in the corruption of the Kremlin and its supporters that they are no longer able to meet the standards expected of a UK-regulated law firm'.[10] It was significant that at the same hearing evidence was received that both the sale of Russian debt in the London market and the Russian En+ Group floatation (which had caused the controversy) 'appear to have been carried out in accordance with the relevant rules and regulatory systems, and there is no evidence of impropriety in a legal sense'.[11]

While the kind of aspersions made against the London law firm are regarded with a degree of cynicism by many in the London financial sector, real anger is felt when the United States attempts to enforce extraterritorial compliance or threatens secondary sanctions. The American public–private co-operation and consultation described above certainly does not extend to any great degree beyond US borders. The withdrawal of the United States from the Joint Comprehensive Plan of Action has substantially increased this resentment. One anonymous European businessman when interviewed by the *New York Times* vented his anger at the Trump administration when he noted: 'The message is this: "Rome and Caesar has changed his mind. If we disobey, our villages will be burned to the ground."' As the author of the newspaper report, Thomas Erdbrink, pointed out: 'Like every other person interviewed for this article, the executive would speak only anonymously, fearing the wrath of either the Iranian or the American authorities.'[12]

This is not to argue that the policy of the American authorities towards Iran does not have value or that European businesses may be putting profit before morality, but it would not be true to suggest that the public–private partnership in the area of finance and security is always a very smooth one.

The private sector is often forced into coalition efforts irrespective of their costs and successes. Again, this is not to argue that the private sector, especially banks and large financial institutions, are not in some respects effectively on the front line of AML and CFT efforts but that sometimes the requirements placed upon them are overwhelming and sometimes not commercially grounded.

Clearly, finance has a role to play in combating terrorist actions, but as terrorism morphs into lone wolf and small group forms, it is arguably the police and community bodies that are increasingly on the front line and more relevant than the banks. In the case of powerful territorial-based organisations such as the Islamic State, it is ground forces not financial institutions that are key, and state authorities cannot be allowed to hide behind financial actions which the private sector, under the threat of fines, must enforce. When it comes to high-end criminal money laundering, it is state authorities that must set the tone and provide the most effective legislative framework in which financial institutions can best bring to bear AML and CFT programmes. Increased spending on government institutions tasked with AML and CFT responsibilities will also help. Such steps will surely be in everyone's interest.

Since the financial crisis of 2008 financial institutions have been subjected to over $320 billion in fines for various AML and sanctions infringements.

Buffeted by these massive costs and enormous penalties, these entities will continue to be forced to act defensively to protect themselves. The defensive actions range from over-reporting of suspicious transactions, which bears the risk of swamping the system, to de-risking, which isolates large areas of the developing world from legitimate global finance and ironically makes those regions more vulnerable to illegitimate funds and exploitation.

It will be very difficult for financial institutions to break out of this pattern. Already in 2005, well before the advent of the big fines for regulatory deficiencies, McKinsey & Co. had identified regulatory fines as a major banking loss event (together with deceptive sales practices, anti-trust settlements, loan fraud and embezzlement). In fact, the McKinsey study demonstrates that monetary penalties associated with these fines are only one type of loss suffered by companies. Other more long-term effects include the associated loss of shareholder value, remediation costs and damaged reputation.[13] Financial institutions need to protect themselves against these monetary and non-monetary penalties and will do so even if the impact on the broader financial system is cumulatively less than desirable.

To increase efficiency and effectiveness, authorities ironically need to work towards reducing and honing relevant compliance demands, not increasing them and burdening hard-pressed compliance departments. Unfortunately, this does not seem to be happening. The City of London, for example, faces an additional compliance headache as a result of Brexit and the need to cope now not only with US, EU (and various national) sanctions systems but also the United Kingdom's post-Brexit regime which, as a result of the Sanctions and Anti-Money Laundering Act 2018, will set its own AML, CFT and sanctions and licensing standards separate from those of the European Union. The one mitigating factor is that there is a degree of discretion in the implementation of the new Act, and one hopes it will be administered with a relatively light touch.

A light touch in administering AML compliance will of course not have as much relevance or benefit in the context of the employment of more offensive financial weapons such as sanctions and arms embargoes, where international co-operation is even more fundamental than in the AML and CFT context.

These financial weapons are set to become increasingly difficult to implement as ideological and political competitors such as China and Russia are able provide alternative sources of financial product and may themselves at some point begin to engage in financial warfare. The emergence of Russian counter-sanctions in the context of the Crimean crisis which at this stage is not particularly significant, arguably adumbrates future developments. Government authorities are also set to become more aware that their own private sector will be increasingly resistant to policies—some of which are not even derived from their own national authorities—that have the consequence of disadvantaging them by freezing them out of competitive markets.

In any event, as former US Treasury Secretary Jack Lew stated in 2016:

> We must be conscious of the risk that overuse of sanctions could undermine our leadership position within the global economy, and the effectiveness of the sanctions themselves. If they make the business environment too complicated or unpredictable, or if they excessively interfere with the flow of funds worldwide, financial transactions may begin to move outside of the United States entirely, which could threaten the central role of the U.S. financial system globally, not to mention the effectiveness of our sanctions in the future.[14]

This is true for the United States and no doubt other jurisdictions as well. Overuse of sanctions and embargoes will simply allow financial and non-financial institutions in other states in Asia and elsewhere to fill the gap vacated by Western institutions.

The United Kingdom has for long played an important role in EU financial sanctions, in terms of designating relevant persons, providing intelligence and driving policy forward. This has benefited both the EU public and private sectors. Following Brexit, it is quite possible that a degree of disjuncture will emerge between the United Kingdom and its former EU partners that even continued co-ordination will not fully resolve. As financial warfare and sanctions are best implemented in the context of a unified front, such a breach could well impact upon the efficacy of this financial weapon.

Even without the break between the United Kingdom and European Union, pressures between the United States and the Europeans will continue to add further instability to the equation. While EU authorities may be able to control their own punitive measures against persons within their jurisdictions, they have far less ability to constrain American decision-making. The danger is not only of over-reporting and de-risking but also that the private sector becomes more inward-looking and less responsive to national foreign policy considerations. While European governments may wish their private sectors to remain committed to doing business with Iran, for example, CEOs will be less impressed with political exhortations than with the dangers of extraterritorial compliance fines and excommunication from the United States market.

Panacea Bombing and Financial Virtue Signalling

> *Had we claimed two years ago to have been able to do half the damage to the German railway system and rolling stock that has since been done, I have not the least doubt that the 'Panacea' mongers would have claimed such a scale of damage as lethal to the entire internal communication system of Germany.*

Head of RAF Bomber Command, Sir Arthur Harris, 20 December 1943[15]

Sir Arthur Harris became head of Bomber Command in February 1942. He believed with certitude and conviction that the most efficient and rapid way to end the war was to target German cities for destruction. Dresden, Hamburg and Cologne were but a few of the cities ravaged by the counter-value bombing operations of the RAF. There were many reasons why the British chose this strategy—technical, political, strategic—and much debate has since raged as to both the effectiveness and morality of this campaign. These controversies do not concern us here. What is of interest is not what Harris desired but what he detested. And these were actions he considered peripheral and irrelevant to what had to be the main weight of the campaign—the destruction of German urban centres. Everything else merely served as a 'panacea'. He derided precision raids on targets such as oil plants, ball-bearing factories, railway depots and even missile sites. He doubted whether the famous Dambusters' raid had achieved anything significant. As for dropping pamphlets on Germany, they merely provided toilet paper to the enemy. All these objectives were what he derided as 'panacea targets' that served more of a psychological—indeed delusional—function.

The argument here is certainly not that finance warfare is a form of panacea warfare (or that Harris was correct in his strategic assessment), but there are nevertheless dangers of financial virtue signalling that need to be recognised. In his seminal study on the financial war against terrorism, Zarate notes explicitly that national authorities and the private sector often do not find it difficult to co-operate against terrorist financing or illicit funding flows because it presents a softer and more manageable form of combating transnational threats than other methods. He notes that for many countries, freezing bank accounts or seizing cash at borders is a more palatable way of fighting terrorism than sending troops to war zones.[16]

While one can debate his reference to the private sector on several levels (other options for countering terrorism are not really available to it, and many within the sector approach the problem with a degree of circumspection), the claim about national authorities and their preference for activities other than military campaigns is indisputably true. When these steps become not a means of increasing leverage and shaping the environment in which actors such as terrorists, money launderers and others function, but rather in effect excuses for non-action in the face of militarily and politically difficult and economically expensive choices, then problems arise.

Too often financial warfare becomes a panacea and part of a range of actions by which states and multilateral organisations can convince themselves and others that they are actually doing something effective rather than actually carrying out the hard decisions such as deploying ground troops or even reaching financial defence expenditure targets. In the final analysis, while we may choose to use the term 'financial warfare', it is nevertheless important to recognise that it is essentially a misnomer. As students of War Studies are well aware, Clausewitz emphasised that warfare is 'an act of violence intended to compel our opponents to fulfil our will'.[17] Financial warfare is essentially a means of avoiding such violence and while it undoubtedly plays a significant role it should not be confused with real warfare or used as an excuse to substitute something else for it, when military or police action is required.

Enhancing Future Capabilities

Despite this pessimistic assessment there are a number of areas where developments may act to enhance defensive AML and CFT capabilities. These are, first, better global and public–private information sharing and, secondly, technological progress.

Potentially, these developments may assist private-sector financial institutions and non-financial institutions in dealing with the central problem they face of ensuring that they meet growing compliance requirements while at the same time managing growing AML, CFT and related costs without significantly impacting upon their revenue and client satisfaction.

We have already touched on the need for better information to be shared between governments and also between institutions in the private sector. More and better targeted information also needs to be supplied by the state to those financial and non-financial institutions active in the private sector. These steps will surely help the private sector navigate very complex compliance commitments and enable them to act far more efficiently. There is little doubt that this is crucial. The Financial Action Task Force has long been explicit in stating: 'Providing accurate and forward-looking guidance to the private sector improves their monitoring and screening processes and reporting-time on sensitive transactions.'[18] This is a view reaffirmed in the British Treasury's AML–CFT supervision report:

> The greatest opportunity to increase effectiveness through 2018 and beyond is sharing—sharing intelligence, skills and experience right across the AML/ [CFT]arena. It is essential that all parts of the regime work together effectively and efficiently—and key to this is developing cross-regime relationships built on trust and shared goals to deter, detect and disrupt the real villains involved in the most serious areas of criminal activity.[19]

Evidence presented to the House of Commons Foreign Affairs Committee on Russian monies entering the UK reached not dissimilar conclusions:

> Some witnesses to this inquiry said that the regulatory architecture for combating money laundering is fairly robust, and that the right laws and frameworks are by and large already in place. However, witnesses also emphasised the need for the Government to dedicate sufficient resources to anti-money laundering (AML) measures, to co-ordinate more effectively, and to demonstrate greater political will in tackling the problem.[20]

Yet cynics cannot avoid noticing that analysts have been pointing out the problems of improved communication and co-ordination, as well as the need for enhanced national global co-operation in this space, for nearly twenty years.[21] Of course, it remains true that such improvements are necessary conditions for more effective 'know your customer' exercises and ultimately the identification and prevention of AML, CFT and other threats to the financial system. Enhanced information sharing at these various levels is nevertheless something far easier said than done. If AML and CFT programmes are to undergo a qualitative improvement, such steps, having not been properly undertaken in the past, will need, despite cross-Atlantic and Brexit tensions, to be undertaken in the future. The relationship between investment in AML resources and results will also need to be established with a greater degree of confidence and precision than in the past.

The second area where AML and CFT capabilities may advance is in the technological space, especially in the field of artificial intelligence, which can efficiently and speedily analyse massive quantities of data in the search for abnormalities.

The present risk-based approach which underpins AML and CFT exhibits major inefficiencies of which perhaps the most striking is the number of false positives that its methodology continues to generate. That is, investigations often give rise to results indicating money laundering or some other illicit activity when this is in fact not the case. These false results consume significant amounts of money and effort. While the high rate of false positives has been reduced with the adoption of enhanced due diligence, their number remains very high. Thus banks have reported that 90 to 99 per cent of all alerts are false positives, and even where the system has been fine-tuned, the figure is still in the range of 80 to 85 per cent. Moreover, a large number of these are multiple alerts which result from the same event or from previously resolved alerts that recur.[22]

As a result some financial institutions are starting to complement rules-based systems with artificial intelligence, unsupervised machine learning (UML) and process automation. As one study has noted: 'Advanced analytical tools have the potential to reveal hidden patterns and risks, streamline alert investigation, automate manual process and continuously learn and improve. They will be essential for improving the efficiency and effectiveness of AML operations.'[23]

Artificial intelligence AML systems have the capability of identifying patterns of money laundering and terrorist financing that are otherwise hidden, while also decreasing the false positive rates and lowering the cost of compliance. UML is a further development in which artificial intelligence software is provided with both the data and the objective but not any expected output, potentially resulting in the identification of as yet unseen data anomalies.

According to one study, when HSBC implemented an artificial intelligence system in the context of AML investigations, the number of investigations decreased by 20 per cent without the number of cases referred for deeper scrutiny being reduced.[24]

Do these technologies hold out a means of resolving the compliance, cost and revenue–client satisfaction dilemma? While military and intelligence experience teaches us that we should be cautious about over-reliance on technology, emerging AI, machine learning and self-improving algorithms represent a potentially revolutionary advance in rendering AML more effective and efficient for both financial institutions and the state.

Joshua Fruth has written that artificial intelligence and UML applications, while enhancing efficiencies by analysing enormous quantities of structured and unstructured data, are not magic bullets for solving money-laundering detection issues. Indeed, artificial intelligence and UML applications 'are largely developed by technical specialists such as computer scientists who are unlikely to possess the requisite law enforcement, intelligence and financial crime backgrounds to effectively target emergent risks.'[25] He argues that AML detection demands consideration of a range of global security matters, public policy issues and regulatory environments, which are all factors not always integrated into artificial intelligence scenarios. The implication of this analysis is that financial institutions will be best served by both technical and more broadly educated compliance personnel.[26] The merger of AI and strategic analysis is an important way forward.

In reviewing the relationship of finance to security and in analysing our assumptions underpinning financial warfare, it is important for us to recognise that finance—as a means of constraining and complicating a target's behaviour—certainly has a role, but it is a more modest one than is sometimes articulated. Strong defensive AML, CFT and counter-proliferation finance systems narrow the scope in which the targeted actors function. It is possible to argue that without these—very expensive—efforts, the money-laundering threat would only escalate and further degrade the integrity of the financial system. Simple cost-benefit analyses when applied to low probability–high impact terrorist scenarios may not always be appropriate. Financial sanctions and embargoes also serve to raise the costs and complicate the decision-making process of targeted states and entities, and in certain situations may have some effect on behaviour even if it is marginal. But in the end, these financial efforts are only subordinate means of warfare that in limited circumstances can support but can never effectively replace other more direct and decisive lines of action vested in national authorities.

NOTES

INTRODUCTION

1. A classic text in this area is Roy Goode, *Goode on Legal Problems of Credit and Security*, Sweet & Maxwell, 4th edn, 2009.
2. SIDA, *Financial Sector Development*, Nov. 2004, pp. 4–5, https://www.sida.se/contentassets/d00d7e9fbc0c4fd7a03b9d9a759b4900/sidas-policy-on-financial-sector-development_1161.pdf.
3. University of Western Ontario, *Financial Centres: What, Where and Why*, https://instruct.uwo.ca/geog/556/Financial%20Centres2.pdf.
4. FATF, United Kingdom, Mutual Evaluation Report, December 2018, pp. 26–27.
5. TheCityUK, *Enabling Growth Across the UK*, https://www.thecityuk.com/assets/2018/Reports-PDF/b408b7220a/Enabling-growth-across-the-UK-2018.pdf, pp. 10–14, (Last accessed, 8 July 2018).
6. US Department of the Treasury, *National Money Laundering Risk Assessment 2015*, p. 3.
7. Neil Irwin, 'What is Glass-Steagall? The 82-year-old banking law that stirred the debate', *New York Times*, 14 Oct. 2015.
8. Brigitte Unger, Melissa Siegel, Joras Ferwerda, Wouter de Kruijf, Madalina Busuioc, Kristen Wokke and Greg Rawlings, *The Amounts and Effects of Money Laundering*, https://www.researchgate.net/publication/46667096_The_Amounts_and_Effects_of_Money_Laundering.
9. The White House, *National Security Strategy*, 1 Feb. 2015.
10. James Clapper, Director of the Office of National Intelligence, 'Worldwide threat assessment of the U.S. intelligence community', Statement for the record, US Senate, Committee on Armed Services, 26 Feb. 2015.
11. Introductory comments by Adam J. Szubin in US Department of the Treasury, *National Terrorist Financing Risk Assessment 2015*.
12. The White House, *National Security Strategy of the United States of America*, Dec. 2017, p. 11.

13. Ibid., p. 12.

14. HM Treasury and Home Office, *National Risk Assessment of Money Laundering and Terrorist Financing*, Oct. 2015, p. 4.

15. HM Treasury and Home Office, *National Risk Assessment of Money Laundering and Terrorist Financing*, Oct. 2017, p. 2.

16. HM Government, *National Security Strategy and Strategic Defence and Security Review*, Nov. 2015, p. 87.

17. HM Treasury and Home Office, *Action Plan for Anti-Money Laundering and Counter-Terrorist Financing*, Apr. 2016, p. 17.

18. Juan C. Zarate, *Treasury's War: The Unleashing of a New Era of Financial Warfare*, Public Affairs, Kindle edn, 2013.

19. Ibid., p. 423.

20. Ibid., pp. 428–9.

21. Lawrence Freedman, *Strategy*, Oxford University Press, 2013, p. 9.

22. HM Treasury and Home Office, *Action Plan for Anti-Money Laundering and Counter-Terrorist Financing*, Apr. 2016, p. 12.

23. Action plan op.cit. Apr. 2016, p. 12.

24. Financial crime: analysis of firms' data, FCA, November 2018, p. 6. The report sought data from over 2000 firms active in the United Kingdom. It noted that the industry employ 11,500 full time equivalent staff within financial crime roles.

25. https://www.fca.org.uk/news/news-stories/2013-fines; https://www.fca.org.uk/news/news-stories/2014-fines; https://www.fca.org.uk/news/news-stories/2015-fines; https://www.fca.org.uk/news/news-stories/2017-fines.

1. THE LAUNDERING OF MONEY

1. FinCEN Advisory, US Department of Treasury, Financial Crimes Enforcement Network, Nov. 1997, Issue 9.

2. Quoted in Amanda Macias, 'Legendary drug lord Pablo Escobar lost $2.1 billion in cash each year—and it didn't matter', *Business Insider UK*, 24 Sept. 2016.

3. Of significance here is the Bank Secrecy Act 1970 which requires US financial institutions to adopt measures to prevent money laundering. Currently, the general rule is that you must file Form 8300, Report of Cash Payments over $10,000 Received in a Trade or Business, if a person's business receives more than $10,000 in cash from one buyer as a consequence of a single transaction or two transactions which are related. See https://www.irs.gov/businesses/small-businesses-self-employed/bank-secrecy-act.

4. FinCEN Advisory, US Department of Treasury, Financial Crimes Enforcement Network, Nov. 1997, Issue 9.

5. Colombian Decree Law 444 (1967) promulgated, inter alia, to protect the peso against wild fluctuations prevented Colombians from retaining foreign currency,

limited the amount of dollars that could be legally purchased from the local banking systems and ensured that foreign goods attracted significant import tariffs or were banned outright.

6. 'Mexico: Organized crime and drug trafficking organisations', Congressional Research Service, 25 April 2017.

7. Beau Kilmer, Jonathan P. Caulkins, Brittany M. Bond and Peter H. Reuter, 'Reducing drug trafficking revenues and violence in Mexico', Rand Occasional Paper, 2010.

8. *The National Money Laundering Risk Assessment, 2015*, US Department of the Treasury, p. 3.

9. *A Line in the Sand: Countering Crime, Violence and Terror at the Southwest Border*, Majority Report of the House Committee on Homeland Security Subcommittee on Oversight, Investigation and Management, Nov. 2012.

10. *The National Money Laundering Risk Assessment 2015*, p. 28.

11. Money may then be re-exported to the US from Mexico as legitimate funds.

12. *The National Money Laundering Risk Assessment 2015*, p. 30.

13. Tim Reason, 'The corporate connection', CFO, http://ww2.cfo.com/risk-compliance/2001/03/the-corporate-connection.

14. 'Estimating illicit financial flows resulting from drug trafficking and other transnational organized crime', UNODC, Oct. 2011.

15. Ibid.

16. UNDOC Money Laundering and Globalization; https://www.unodc.org/unodc/en/money-laundering/globalization.html (Last Accessed 2 January 2019).

17. *National Strategic Assessment of Serious and Organised Crime 2016*, NCA, Sept. 2016, p. 28.

18. Evidence before House of Commons Home Affairs Committee, Proceeds of Crime, Fifth Report of Session 2016–2017, 15 July 2016, p. 21.

19. See a brief history in International Money Laundering Information Bureau, http://www.imlib.org/home.html.

20. The expression first appeared in a legal context in 1982 in the US courts in the case *US v $4,255,625.39 (1982)* 551 F Supp. 314.

21. *UK National Risk Assessment of Money Laundering and Terrorist Financing 2017*, p. 19.

22. Jonathan Hoey, 'The risk of third party payments: How to avoid money laundering and other financial crime', TLT, 27 July 2016.

23. ICE, HSI, Trade Transparency Unit overview of TBML, http://www.ice.gov/trade-transparency.

24. Ibid.

25. For an argument as to the limitations of the PLI Model, see Paul Camacho, 'Whales, sharks and flounders: Conceptualizing real world money laundering', ACFCS, 25 Jan. 2018, https://www.acfcs.org/news/383891/WHALES-

SHARKS-AND-FLOUNDERS-CONCEPTUALIZING-REAL-WORLD-MONEY-LAUNDERING.htm.

26. *National Money Laundering Risk Assessment 2015*, p. 37.
27. Ibid., pp. 38–9.
28. ACAMS, *Risks and Methods of Money Laundering and Terrorist Finance*, ACAMs, ch. 2, p. 16, http://files.acams.org/pdfs/English_Study_Guide/Chapter_2.pdf.
29. 'A POS System or point of sale system is the term used to describe a system of software and hardware that aid retailers and restaurants with sales, inventory management, analytics, reporting and employee management. A typical POS System consists of a computer that runs the POS software, a receipt printer, barcode scanner, credit card reader and an on-site server for legacy based point of sale systems.' See https://www.lightspeedhq.com/blog/2015/06/what-is-a-pos-purchase-and-other-pos-term-clarifications.
30. Qian Sun, p. 7.
31. Henry McDonald, 'Criminal gangs using high-street nail bars to launder dirty money', *The Guardian*, 19 Sept. 2006.
32. 'Offshore activities and money laundering: Recent findings and challenges', Study for the PANA Committee, Policy Department, European Parliament, March 2017, p. 42.
33. Data on more than 290,000 offshore companies from the Paradise Papers has now been released. See https://www.icij.org/.
34. Scott Shane, 'Panama Papers reveal wide use of shell companies by African officials', *New York Times*, 25 July 2016.
35. See https://www.gov.uk/government/publications/guidance-to-the-people-with-significant-control-requirements-for-companies-and-limited-liability-partnerships.
36. Comments by Keith Vaz, MP, 'End London "welcome mat" for money launders says committee', www.parliament.uk, 15 July 2016.
37. House of Commons, Home Affairs Committee, Proceeds of Crime, Fifth Report of Session 2016–17, 15 July 2016, p. 21.
38. 'London Property: A top destination for money launderers', Transparency International, Dec. 2016, p. 4.
39. Ibid.
40. Andy Verity and Nassos Stylianou, 'Firms on Caribbean island chain own 23,000 UK properties', www.bbc.co.uk/news/business-42666274.
41. The British Virgin Islands argues that there are many reasons why properties may be owned by locally incorporated entities including legal structures which enable the combination of multiple investors often necessary for large commercial property transactions.
42. Judith Evans, 'How laundered money shapes London's property market', *Financial Times*, 6 April 2016.

43. See Haines Watts, 'Money laundering: High value dealers', Haines Watts Fact Sheet, 2017, https://www.hwca.com/resources/factsheets/money-laundering-high-value-dealers/.

44. *UK National Risk Assessment 2017*, p. 81.

45. *The Queen on the Application of Mohammed Sani Abacha, Abubakar Bagdu v The Secretary of State for the Home Department v the Federal Republic of Nigeria*, Queens Bench Division, [2001] EWHC Admin 787, 2001 WL 1171852.

46. United Nations Convention against Corruption (UNCAC), UNODC, 2004, p. 19.

47. See 'How do you define corruption', Transparency International, https://www.transparency.org/what-is-corruption#define. There is an extensive literature on the broader impact of corruption and bribery on the state. Two other main types of corruption are political corruption involving the abuse by decision-makers of policies and procedures to benefit themselves; and petty corruption involving mid- and low-level public officials who abuse their positions when interacting with the public attempting to access goods and services.

48. https://www.transparency.org/news/feature/corruption_perceptions_index_2017.

49. See for example *Wolfsberg Guidance on Politically Exposed Persons*, The Wolfsberg Group, 2017.

50. 'Banks' management of high money-laundering risk situations: How banks deal with high-risk customers (including politically exposed persons), correspondent banking relationships and wire transfers', FSA, June 2011.

51. *Moscow's Gold: Russian Corruption in the UK*, House of Commons Foreign Affairs Committee, Eighth Report of Session, 2017–2019, p. 4.

52. Ibid., p. 18.

53. Ibid., p. 10.

54. *NRA 2017*, p. 29

55. *NRA 2017*, p. 30.

56. Jeffrey Grocott and Gregory White, 'How "mirror trades" moved billions from Russia: Quick take Q&A', *Bloomberg Politics*, 28 June 2017.

57. Ed Caesar, 'Deutsche Bank's $10-billion scandal: How a scheme to help Russians secretly funnel money offshore unravelled', *The New Yorker*, 9 Aug. 2016.

58. The Global Laundromat information obtained by the Organized Crime and Corruption Reporting Project (OCCRP) and *Novaya Gazeta*. OCCRP shared the data with the *Guardian* and media partners in 32 countries. https://www.occrp.org/en.

59. Luke Harding, Nick Hopkins and Caelainn Barr, 'British banks handled vast sums of laundered Russian money', *The Guardian*, 20 March 2017.

60. See Accountancy Affinity Group's supervisory pages, www.Accountancysupervisors.co.uk; also see CIMA, 'Anti Money Laundering' at CIMAglobal.com; also see

Rosie Murray-West, 'How accountants can prevent criminal activities', *The Telegraph*, 10 Feb. 2017.

61. *NRA 2017*, p. 45.
62. Solicitors Regulation Authority Risk Outlook 2017/18 July 2017, p. 34.
63. *NRA 2017*, p. 54.
64. *Professional Money Laundering*, FATF, July 2018, pp. 10–12.
65. Ibid., p. 14.

2. THE FINANCING OF TERROR

1. Shan Carter and Amanda Cox, 'One 9/11 tally $3.3 trillion', *New York Times*, 8 Sept. 2011.
2. In November 2017 Bakr was arrested along with many others in a purge conducted by Crown Prince Mohammed bin Salman. See www.sbg.com.sa/profile.html.
3. John K. Cooley, *Unholy Wars*, Pluto Press, 2000, p. 118.
4. www.sbg.com.sa/profile.html.
5. *The 9/11 Commission Report: The Report of the National Commission on Terrorist Attacks upon the United States*, W.W. Norton, p. 497 n. 112.
6. Ibid., pp. 498 n. 126 and 499 n. 127.
7. Peter L. Bergen, *Holy War*, Phoenix, 2002 p. 119.
8. *9/11 Commission Report*, p. 55.
9. Ibid., p. 56.
10. Ibid., p. 57.
11. Ibid., p. 170.
12. Daniel Benjamin and Steven Simon, *The Age of Sacred Terror*, Random House, 2003, p. 111.
13. Ibid., p. 111.
14. Lawrence Wright, *The Looming Tower*, Knopf, 2006, p. 196.
15. *9/11 Commission Report*, p. 58.
16. Ibid., pp. 57, 170.
17. Wright, *The Looming Tower*, p. 196.
18. *9/11 Commission Report*, p. 66.
19. Richard A. Clarke, *Against All Enemies*, Simon & Schuster, 2004, p. 195.
20. *9/11 Commission Report*, p. 171.
21. Ibid., p. 498 n. 114.
22. Ibid., p. 499 n. 130.
23. Ibid., p. 499 n. 130.
24. Ibid., p. 172.
25. Ibid., p. 224.
26. The *9/11 Commission Report* states that the so-called twentieth hijacker Zacharias Moussaoui spent $50,000 though it is unclear if the report is stating that this was

in addition to the $270,000. The report also says that the $270,000 was spent within the United States but, when broken down, the expenses appear to include those incurred outside the United States as well (p. 499 n. 131).

27. Ibid.
28. Ibid. Part Four: Finding, Discussion and Narrative Regarding Certain Sensitive National Security Matters, p. 416. Report contained in David Smith and Spencer Ackerman, '9/11 reports classified "28 pages" about potential Saudi Arabia ties released', *The Guardian*, 15 July 2016.
29. Ibid.
30. Simon Henderson, 'What we know about Saudi Arabia's role in 9/11', *Foreign Policy*, 18 July 2016.
31. *Consolidated FATF Strategy on Combating Terrorist Financing*, FATF, Feb. 2016.
32. Emile Oftedal, 'The financing of jihadi terrorist cells in Europe', Norwegian Defence Research Establishment (FFI), Norway, 6 Jan. 2015.
33. *Terrorist Financing in West Africa*, FATF, Oct. 2013, pp. 12–15.
34. See case studies in *Financial Flows Linked to the Production and Trafficking of Afghan Opiates*, FATF, June 2014.
35. See for example, *United States National Terrorist Financing Risk Assessment 2015*, US Department of Treasury.
36. 'Financial flows linked to the production and trafficking of Afghan opiates', p. 43.
37. *National Terrorist Financing Risk Assessment 2015*, p. 14.
38. *Organised Maritime Piracy and Related Kidnapping for Ransom*, FATF, July 2011, p. 29.
39. Robert F. Worth, 'How a ransom for royal falconers reshaped the Middle East', *New York Times* Magazine, 14 March 2018.
40. David Cohen remarks before the Center for a New American Security, 4 March 2014, https://www.treasury.gov/press-center/press-releases/Pages/jl2308.aspx.
41. *National Terrorist Financing Risk Assessment 2015*, p. 15.
42. *Outcomes FATF Plenary, 21–23 February 2018*, FATF, Feb. 2018.
43. *UK National Risk Assessment of Money Laundering and Terrorist Financing 2017* (NRA 2017), pp. 26–8.
44. John Rollins and Wyler Liana Sun, 'Terrorism and transnational crime: Foreign policy issues for Congress', Congressional Research Service, 11 June 2013.
45. See Tamara Makarenko, 'The crime-terror continuum: Tracing the interplay between transnational organized crime and terrorism', *Global Crime*, Vol. 6, No. 1, Feb. 2004.
46. UK *NRA 2017*, p. 28.
47. See 'Transnational organized crime in West Africa: A threat assessment', UN Office on Drugs and Crime, 25 Feb. 2013.
48. The list is determined by three laws: Section 6j of the Export Administration Act, Section 40 of the Arms Export Control Act and Section 620 of the Foreign Assistance Act.

49. *Country Reports on Terrorism 2015*, U.S. Department of State, June 2016.

50. *Country Reports on Terrorism 2016*, U.S. Department of State, July 2017.

51. Adam Taylor, 'North Korea's on-again-off-again status as a state sponsor of terrorism', *Washington Post*, 20 Nov. 2017.

52. Jonathan Masters and Mohammed Aly Sergie, 'Al-Shabaab', *Council on Foreign Relations Backgrounders*, 5 March 2015.

53. *UK Mutual Evaluation Report*, FATF Dec. 2018, p. 100.

54. David Cohen remarks before the Center for a New American Security, 4 March 2014.

55. David Smith and Spencer Ackerman, '9/11 reports classified "28 pages" about potential Saudi Arabia ties released', *The Guardian*, 15 July 2016.

56. *National Terrorist Financing Risk Assessment 2015*, p. 17.

57. *Risk of Terrorist Abuse in Non-Profit Organisations*, FATF, June 2014, ch. 3.

58. 'Terror trial: Public duped into funding bomb plotters', BBC, 23 Oct. 2012, https://www.bbc.co.uk/news/uk-20040835 Last Accessed 18 July 2018.

59. *Risk of Terrorist Abuse in Non-Profit Organisations*, FATF, June 2014, p. 44.

60. Ibid. p. 46.

61. *UK Mutual Evaluation Report*, op. cit., p. 89.

62. Oftedal, 'The financing of jihadi terrorist cells in Europe', p. 7.

63. Duncan Gardham, 'Terror trail: The doctors accused of planning "spectacular" car bombing', *The Telegraph*, 9 Oct. 2008.

64. *UK Mutual Evaluation Report*, op. cit., p. 50.

65. Ibid. p. 88.

66. *Emerging Terrorist Financing Risks*, FATF, Oct. 2015, pp. 24–42.

67. Ibid., p. 31.

68. Ibid., p. 30.

69. 'Foreign fighters', *The Soufan Group*, Dec. 2015. Also see Tom Keatinge, 'Identifying foreign terrorist fighters: The role of public-private partnership, information sharing and financial intelligence', Royal United Services Institute, Aug. 2015.

70. The FATF study also drew attention to the exploitation of new payment methods which will be addressed in Chapter 3.

71. This section will eschew discussion of what constitutes a terrorist organisation and the meaning of statehood (for example whether ISIS was a state or a proto-state) but will rather focus on the financial elements of the Islamic State.

72. S. Heibner, P.S. Neumann, J. Holland-McCowan and R. Basra, 'Caliphate in decline: An estimate of Islamic State's financial fortunes', ISCR and EY, 2017, p. 3.

73. Emily Glazer, Nour Malas and John Hisenrath, 'US cut cash to Iraq on Iran, ISIS fears', *Wall Street Journal*, 3 Nov. 2015.

74. Heissner et al., 'Caliphate in decline', p. 8.

75. ibid., p. 8.

76. Ibid., p. 7.

77. For a discussion of taxes see Carla Humud, Robert Pirog and Liana Rosen, 'Islamic State financing and US policy approaches', Congressional Research Service, April 2015, pp. 9–10.

78. *Emerging Terrorist Financing Risks*, FATF, Oct. 2015, p. 40.

79. The Global Coalition—Working to Defeat ISIS, US Department of State, Fact Sheet, 22 Mar. 2017.

80. Ibid. It was claimed that these operations ended in 'destroying tens of millions- and possibly hundreds of millions of dollars'.

81. Ibid.

82. 'The financing of the Islamic State in Iraq and Syria', Directorate-General for External Policies, Policy Department, Sept. 2017, p. 4.

83. The report claimed for example that the ISIS position in Iraq and Syria may be weakening 'in particular due to lack of resources'. Ibid., p. 18.

84. Matthew Levitt, 'Hezbollah finances, funding the Party of God', The Washington Institute, Feb. 2005.

85. Josh Meyer, 'The secret backstory of how Obama let Hezbollah off the hook', *Politico*, 17 Dec. 2017. The report details Hezbollah's involvement in drug smuggling and money laundering. More controversially, it alleges that the Obama administration soft-pedalled the investigation into these Hezbollah activities in order to save the nuclear deal with Iran, something former officials of the administration strenuously deny.

3. THE TRANSFER OF ILLICIT FUNDS

1. Hugh Son, Hannah Levitt and Brian Louis, 'Jamie Dimon slams Bitcoin as a "fraud"', *Bloomberg Technology*, 12 Sept. 2017. Dimon later said that he regretted calling Bitcoin a fraud.

2. In the United States financial institutions must in respect of cash transactions with a customer during a single business day in excess of $10,000 (whether by way of a single transaction or a series of related transactions) file a currency transaction report with FinCEN, while non-financial businesses must report to FinCEN and the IRS in relation to cash transactions exceeding $10,000.

3. UK Mutual Evaluation Report, FATF Dec. 2018, p. 52.

4. *Money Laundering through Physical Transport of Cash*, FATF, Oct. 2015, pp. 3–5.

5. *National Terrorist Financing Risk Assessment 2015*, United States Department of Treasury. p. 54.

6. Ibid., p. 47.

7. Ibid., p. 55.

8. *United States v. Hor and Amera Akl*, No. 3:10-cr-00251-JGC (N.D. Ohio, filed 7 June 2010).

9. *National Terrorist Financing Risk Assessment 2015*, pp. 55–6.

10. *UK National Risk Assessment of Money Laundering and Terrorist Financing 2017* (*NRA 2017*), HM Treasury and Home Office, p. 67.

11. Ibid.

12. Ibid.

13. *Terrorist Financing February 2008*, FATF, Oct. 2008, p. 24.

14. Phil Stewart and Lesley Wroughton, 'How Boko Haram is beating U.S. efforts to choke its financing,' Reuters, 1 July 2014, http://www.reuters.com/article/2014/07/01/us-usanigeria-bokoharam-insight-idUSKBN0F636920140701.

15. *National Terrorist Financing Risk Assessment 2015*, pp. 54–6.

16. *Emerging Terrorist Financing Risks*, FATF, Oct. 2015.

17. *Terrorist Financing*, FATF, p. 22–3.

18. *Emerging Terrorist Financing Risks*, FATF, p. 21.

19. Peter R. Neumann, 'Don't follow the money', *Foreign Affairs*, Vol. 96, No. 4, July/Aug. 2017, p. 95.

20. *National Terrorist Financing Risk Assessment 2015*, pp. 49–50.

21. 'Banks' management of high money-laundering risk situations', Financial Services Authority, June 2011, p. 44.

22. Ibid., p. 49.

23. Ibid., p. 52.

24. Ibid., p. 52.

25. Ibid., p. 53.

26. Ibid., p. 59.

27. In the United States the Federal Financial Institutions Examination Council requires that any United States bank maintaining a relationship with a foreign correspondent institution in an overseas jurisdiction must set out the terms of that relationship in an explicit contract.

28. 'Foreign correspondent banking: The good, the bad and the ugly', Dinsmore & Shohl LLP, 11 Feb. 2016.

29. FATF Public Statement, FATF, 19 Feb. 2016.

30. See *FATF Guidance: Correspondent Banking Services*, Oct. 2016.

31. See for example Basel Committee on Banking Supervision, *Guidance on Sound Management of Risks Related to Money Laundering and Financing of Terrorism*, 2014, The Wolfsberg Group, *Anti-Money Laundering Principles for Correspondent Banking*, 2014; The Wolfsberg Group, *Anti-Money Laundering Questionnaire*, 2014; Committee on Payments and Market Infrastructures, *Correspondent Banking: Consultative Report*, 2015; Basel Committee on Banking Supervision, *Supervisory Guidance for Managing Risks Associated with the Settlement of Foreign Exchange Transactions*, 2016.

32. Neil Buckley, 'Latvia, a Banking Scandal on the Baltic', *Financial Times*, 23 Feb. 2018.

33. Richard Milne, 'Latvia banks still complicit in money laundering, claims US', *Financial Times*, 8 March 2018.

34. 'A crackdown on financial crime means global banks are de-risking', *The Economist*, 8 July 2017.

35. The Financial Stability Board attempts to co-ordinate national financial authorities and international standard-setting bodies so as to promote and co-ordinate regulatory, supervisory and other financial sector policies.

36. 'A crackdown on financial crime means global banks are de-risking', *The Economist*, 8 July 2017.

37. Financial Stability Board, *Report to the G20 on Actions taken to Assess and Address the Decline in Correspondent Banking*, 6 Nov. 2015.

38. Neumann, 'Don't follow the money', p. 96.

39. J.C. Zarate, *Treasury's War*, Public Affairs, 2013, p. 424.

40. See *FATF Guidance: Correspondent Banking Services*.

41. Terms often interchangeably used in respect of the informal banking include alternative remittance systems, money service businesses, money value transfer systems, money transmitters and Hawalas. Also see the terms 'hundi', 'fei ch 'ien', 'chit system', 'poey kuan'. FATF uses the term HOSSP—Hawala and other similar service providers.

42. This is not always the case and records are in some instances kept though obviously not if the intention is illegal.

43. United States Department of Treasury, *Hawalas and Alternative Remittance Systems*, https://www.treasury.gov/resource-center/terrorist-illicit-finance/Pages/Hawala-and-Alternatives.aspx.

44. *The Role of Hawala and Other Similar Service Providers in Money Laundering and Terrorist Financing*, FATF, Oct. 2013, p. 25.

45. Ibid., p. 14.

46. Ibid., p. 41.

47. Ibid., pp. 41–3.

48. *Migration and Remittances Factbook*, World Bank, 2016.

49. *NRA 2017*, p. 68.

50. Ibid., p. 69.

51. Ibid., p. 70.

52. Ibid.

53. In the United Kingdom Her Majesty's Revenue and Customs (HMRC) supervises money service businesses while the Financial Conduct Authority (FCA) supervises those firms regulated by the FCA under the Financial Services and Markets Act 2000. A Proceeds of Crime Intervention Team located in HMRC focuses on money service business money laundering.

54. *National Terrorist Financing Risk Assessment 2015*, p. 51.

55. See FinCEN Advisory, *Informal Value Transfer Systems*, FIN-2010-A011, 1 Sept. 2010, https://www.fincen.gov/resources/advisories/fincen-advisory-fin-2010-a011.

56. *National Terrorist Financing Risk Assessment 2015*, p. 52.

57. 'Lessons learned from the Paris and Brussels terrorist attacks', *ACAMS Today*, 29 March 2016, https://www.acamstoday.org/lessons-learned-paris-brussels-attacks/.

58. Mark Maremont and Christopher S. Stewart, 'FBI says ISIS used eBay to send terror cash to the US', *Wall Street Journal*, 10 Aug. 2017.

59. 'Emerging terrorist financing risks', p. 38.

60. Of course, had they then held onto their Bitcoin investments into 2018 those gains would have been lost.

61. *Virtual Currencies: Key Definitions and Potential AML/CFT Risks*, FATF, 2014; and *Guidance to a Risk-Based Approach to Virtual Currencies*, FATF, 2015.

62. Joshua Bearman, Tomer Hanuka, Joshua Davis and Steven Leckart, 'The rise and fall of Silk Road', Parts 1 and 2, *Wired*, Jan. 2012.

63. See comments by M. McGuire in 'Virtual cash helps cyber-thieves launder money research suggests', BBC, 18 March 2019, which refers to P. Dreyer et al., 'Estimating the global cost of cyber risk', Rand, 15 Jan. 2018.

64. *Internet Organised Crime Threat Assessment 2017*, Europol, p. 13.

65. 'Up to S200 billion in illegal cybercrime profits is laundered each year, comprehensive research study reveals', Bromium Press Release, 16 March 2018.

66. *National Strategic Assessment of Serious and Organised Crime 2016*, National Crime Agency, 9 Sept. 2016, p. 29.

67. 'Emerging terrorist financing risks', p. 35.

68. 'Former CIA analyst say terrorists utilize Bitcoin to boost funds', CCN, 30 Aug. 2016, https://www.ccn.com/former-cia-analyst-says-terrorists-utilize-bitcoin-boost-funding/.

69. For how Blockchain can be used as an AML tool, see Floyd Costa, 'Blockchain for AML', *International Banker*, 10 Nov. 2017, https://internationalbanker.com/technology/blockchain-aml-harnessing-blockchain-technology-detect-prevent-money-laundering/.

70. David Carlisle, 'Cryptocurrencies and terrorist financing: A risk, but hold the panic', RUSI, 2 March 2017.

71. Ibid. Also see *Guidance for a Risk Based Approach to Virtual Currencies*.

72. UK Mutual Evaluation Report, op. cit. Dec. 2018, p. 201.

4. GLOBAL RESPONSES AND THE RISK-BASED APPROACH

1. *25 Years and Beyond*, FATF, 2014, p. 18.

2. Mike Corder, 'Ukraine files case against Russia at UN's highest court', *US News & World Report*, 17 Jan. 2017.

3. UNODC Treaties, https://www.unodc.org/unodc/en/treaties/index.html.

4. Ibid.

5. For a full list, see United Nations Office of Counter-Terrorism, International Legal Instruments, http://www.un.org/en/counterterrorism/legal-instruments.shtml.

6. International Convention for the Suppression of the Financing of Terrorism, General Assembly Resolution 54/109, 9 Dec. 1999.

7. UNSCR 1267 (1999), http://www.un.org/ga/search/view_doc.asp?symbol=S/RES/1267 (1999).

8. UNSCR 1617 (2005), http://www.un.org/ga/search/view_doc.asp?symbol=S/RES/1617%20%282005%29.

9. The United Nations Global Counter-Terrorism Strategy, General Assembly Resolution 60/288, 8 Sept. 2006. The strategy is reviewed every two years. See ttps://www.un.org/counterterrorism/ctitf/en/un-global-counter-terrorism-strategy.

10. Ibid.

11. Ibid.

12. The CTC is supported by the Counter-Terrorism Committee Executive Directorate (CTED), which carries out the policy decisions of the committee as well as conducting assessments of United Nations members. It provides counter-terrorism technical assistance to states. Also see UNSCR 1535 (2004) which established the CTED, UNSCR 1624 (2005) prohibiting incitement to commit terrorist attacks, and UNSCR 2178 (2014) on countering foreign terrorist fighters and violent extremism.

13. See United Nations Office of Counter-Terrorism, Counter-Terrorism Implementation Task-Force, https://www.un.org/counterterrorism/ctitf/en.

14. For more information on these activities, see UNSCR 1267 (1999), 1989 (2011) and 2253 (2011).

15. See UNSCR 1540 (2004).

16. *Activities of the United Nations System in Implementing the United Nations Global Counter-Terrorism Strategy*, Report of the Secretary General, A/70/826, 12 April 2016, https://www.un.org/counterterrorism/ctitf/en/report-activities-united-nations-system-implementing-united-nations-global-counter-terrorism.

17. United Nations Convention against Corruption, UNODC. Sometimes referred to as the Merida Convention, UNCAC was adopted by the United Nations General Assembly in 2003 and entered force in December 2005, https://www.unodc.org/pdf/crime/convention_corruption/signing/Convention-e.pdf.

18. Also see, for example, the OECD Convention on Combating Bribery of Foreign Public Officials in International Business Transactions which, inter alia, also requires state parties to make the payment of bribes an offence. At the European level, also see the Criminal Law Conventions on Corruption (1999 and 2003) and the Civil Law Convention on Corruption (1999). The Group of States against Corruption monitors implementation of the conventions. In addition there is the Global Infrastructure Anti-Corruption Centre, which is an international organisation providing anti-corruption resources and services focusing on the infrastructure, construction and engineering sectors. Transparency International, a

non-profit organisation, monitors countries, engages with business and provides information with the goal of reducing bribery and corruption.

19. 'Challenges and opportunities for the United Nations', Antonio Guterres, 4 April 2016, https://www.un.org/pga/70/wp-content/uploads/sites/10/2016/01/4-April_Secretary-General-Election-Vision-Statement_Portugal-4-April–20161.pdf.

20. These include the African Development Bank, the Anti-Money Laundering Liaison Committee of the Franc Zone, the Asian Development Bank, the Basel Committee on Banking Supervision, the European Bank for Reconstruction and Development, the European Central Bank, Eurojust, Europol, Group of International Finance Centre Supervisors, the Organisation of American States Inter-American Committee Against Terrorism and the Inter-American Drug Abuse Control Commission, the Organisation for Economic Co-operation and Development, the Organisation for Security and Co-operation in Europe, and the World Customs Organization. These are all FATF observer organisations.

21. Relevant regional groupings here are Asia-Pacific, Caribbean, Eurasia, Eastern and Southern Africa, Latin America, West Africa, Middle East and North Africa, and Central Africa. Also included in this context is the Council of Europe Committee of Experts on the Evaluation of Anti-Money Laundering Measures and the Financing of Terrorism.

22. *25 Years and Beyond*, p. 2.

23. *International Standards on Combating Money Laundering and the Financing of Terrorism & Proliferation: The FATF Recommendations*, FATF, Feb. 2012.

24. Ibid., pp. 11–30.

25. 'The risk based approach: What is it?', The Law Society, 10 Sep. 2008, https://www.lawsociety.org.uk/support-services/advice/articles/what-is-risk-based-approach/.

26. '23,000 people have been "subjects of interest" as scale of terror threat emerges after Manchester attack', *The Telegraph*, 27 May 2017.

27. FATF, *International Standards*, p. 11.

28. Ibid.

29. Ibid.

30. Florentin Blanc and Ernesto Franco-Temple, 'Introducing a risk-based approach to regulate businesses: How to build a risk matrix to classify enterprises or activities', The World Bank, Documents and Reports, 1 Sept. 2013, http://documents.worldbank.org/curated/en/102431468152704305/pdf/907540BRI0Box30d0approach0Sept02013.pdf.

31. *Guidance on the Risk-Based Approach to Combating Money Laundering and Terrorist Financing: High Level Principles and Procedures*, FATF, June 2007.

32. *National Money Laundering and Terrorist Financing Risk Assessment*, FATF, Feb. 2013, p. 21.

33. See *UK National Risk Assessment of Money Laundering and Terrorist Financing*, HM Treasury and Home Office, Oct. 2015.

34. See *UK National Risk Assessment of Money Laundering and Terrorist Financing*, HM Treasury and Home Office, Oct. 2017.

35. See *National Money Laundering Risk Assessment*, United States Department of the Treasury, 2015 and *National Terrorist Financing Risk Assessment*, United States Department of the Treasury, 2015.

36. FATF, *Guidance on the Risk Based Approach*, p. 22.

37. See, for example, the Wolfsberg Statement, *Guidance on a Risk-Based Approach for Managing Money Laundering Risks*, 2006, https://www.wolfsberg-principles.com/sites/default/files/wb/pdfs/wolfsbergstandards/15.%20Wolfsberg_RBA_Guidance_%282006%29.pdf.

38. Transparency International would be one such credible source.

39. Private companies have developed packages of risk products and services that support a firm's operation of an RBA. See, for example, Thomson Reuters Org ID, which provides a KYC-managed service that supports risk identification based on identity data, and documents and conducts ongoing monitoring, so as to alert a firm as to any changes in relation to a corporate customer and the risks that may result.

40. FATF, *International Standards*, p. 21. Also see the Basel Committee on Banking Supervision, *Guidelines: Sound Management of Risks Related to Money Laundering and Financing of Terrorism*, Feb. 2016, which aims to support FATF recommendations by setting out guidelines as to how banks should integrate RBA within their broader risk management framework.

41. *Guidance for a Risk-Based Approach: The Banking Sector*, FATF, Oct. 2014. pp. 12–26.

42. Ibid.

43. Ibid., pp. 27–44.

44. *Financial Crime: A Guide for Firms*, The Financial Conduct Authority, April 2015, https://www.handbook.fca.org.uk/handbook/document/FC1_FCA_20150427.pdf.

45. *Guidance for a Risk-Based Approach: The Banking Sector*, Annex 1.

46. FATF Methodology For Assessing Technical Compliance with the FATF Recommendations and Effectives of AML/CFT Systems, *FATF*, Updated Feb. 2018, p. 15–16.

47. *Anti-Money Laundering and Counter-Terrorist Financing Measures: Iceland. Mutual Evaluation Report*, FATF, April 2018, p. 6.

48. See http://www.fatf-gafi.org/countries/#high-risk.

49. *Outcomes FATF-MENAFATF Joint Plenary, 27–29 June 2018*, FATF, June 2018.

50. The G20's Anti-Corruption Working Group co-ordinates closely with FATF on anti-corruption measures.

51. *The Use of the FATF Recommendations to Combat Corruption*, FATF, Oct. 2013. The following FATF documents also have a particular anti-corruption focus: *A*

Reference Guide and Information Note on the Use of the FATF Recommendations to Support the Fight against Corruption (the Corruption Information Note) (2010 and updated in 2012); *Guidance on Politically Exposed Persons* (Recommendations 12 and 22) (2013); *Specific Risk Factors in Laundering the Proceeds of Corruption: Assistance to Reporting Entities* (2012); *Laundering the Proceeds of Corruption* (2011).

52. *The Use of the FATF Recommendations to Combat Corruption*, FATF, Oct. 2013, p. 22.

53. *Outcomes FATF Plenary, 21–23 February 2018*, FATF, Feb. 2018.

5. NATIONAL ANTI-MONEY LAUNDERING AND COUNTER-TERRORIST FINANCING ARCHITECTURES

1. David Aufhauser, General Counsel, written testimony before the House Financial Services Committee Subcommittee on Oversight and Investigations, 24 Sept. 2003, https://www.treasury.gov/press-center/press-releases/Pages/js758.aspx.

2. *Jayesh Shah, Shaleetha Mahabeer v HSBC Private Bank (UK) Limited*, High Court of Justice Queen's Bench Division, [2012] EWHC 1283 (QB), 16 May 2012. Specifically Supperstone J found for the defendant on the basis of an implied term in the contract which allowed the defendant not to carry out a received payment instruction without an appropriate consent under section 335 of POCA, in circumstances where it suspected the transaction was connected to the proceeds of crime.

3. UK Mutual Evaluation Report, FATF, Dec. 2018, p. 5.

4. The Fifth Anti-Money Laundering Directive focuses on due diligence payments and transparency of beneficial ownership. European Union member states have until January 2010 to implement it in national law.

5. Earlier, in England and Wales the Criminal Justice Act 1988 (as amended by the Criminal Justice Act 1993) made it an offence to assist in the retention or use of the proceeds of serious criminal conduct. Also see the Drug Trafficking Act 1994. Also relevant here is the Fraud Act 2006.

6. Primary legislation refers to Acts of Parliament or statute. Secondary legislation, sometimes described as delegated legislation, is where additional law-making powers are granted to another branch of government by an Act or statute.

7. These include for example the various amendments to the legislation described in the text as well as the Serious Crime Act 2007, the Crime and Court Act 2013, the Fraud Act 2006, the Criminal Finances Act 2017, the Sanctions and Anti Money Launder Act 2018.

8. POCA 2002 also provides financial investigatory powers to the police, the National Crime Agency and customs authorities. It also sets out the legislative framework enabling the recovery of criminal assets.

9. See section below on The Supervision of Compliance in the Regulated Sector.

10. Note that the principal offences in POCA 2002 (as well as those in TA 2000 and the Fraud Act 2006) apply to both the regulated and the unregulated sector. The Money Laundering regulations apply to the regulated sector as do the secondary offences in POCA 2002.

11. Jonathan Peddie, 'Anti-terrorism legislation' in *Banks and Financial Crime*, edited by William Blair, Richard Brent and Tom Gran, Tom, OUP, 2017, p. 422.

12. See CONTEST, the United Kingdom's Strategy for Countering Terrorism, HM Government, July 2011.

13. Both the principal terrorist financing and money laundering offences under the Terrorism Act 2000 and POCA respectively may be punishable by up to 14 years' imprisonment.

14. These orders are issued by the Treasury and not a court. They are in the form of statutory instruments and need to be approved by a resolution of both Houses of Parliament otherwise they cease having effect after 28 days of being made.

15. Andrey Lugovoy and Dmitri Kovtun Freezing Order, 2016, SI2016/67.

16. He was a specifically designated person under the Al-Qaeda and Taliban (United Nations Measures) Order 2006.

17. See the Al-Qaeda and Taliban (United Nations Measures) Order 2002.

18. *HM Treasury v Ahmed* [2010] UKSC 2, [2010] 2 WLR 378.

19. Orders against banks may take the form of Production Orders (POCA 2002); Customer Information Orders (POCA 2002), Account Monitoring Orders (POCA 2002 and TA 2000); Freezing Orders (ATCSA 2001); and Search and Seizure Warrants, Confiscation and Restraint (POCA 2002 and ATCSA 2001).

20. The MLR 2017 came into force on 26 June 2017 and implements the Fourth Money Laundering Directive ((EU) 2015/849) (4MLD) and the Funds Transfer Regulation ((EU) 2015/847).

21. The Treasury also approves AML and CFT guidance produced by many industry sectors covering the application of regulatory and legal requirements for their businesses.

22. Ibid.

23. Ibid.

24. The UK Mutual Evaluation Report, FATF, Dec. 2018, p. 43.

25. Ibid. pp 36–37.

26. For a more detailed analysis of the roles of each of these organisations, ibid. p. 23–27.

27. The NCA was established by the Crime and Courts Act 2013. The powers of the NCA may be extended to counter-terrorism functions but at the time of writing no orders have been made. Nevertheless the NCA certainly functions within the counter-terrorism space in terms of the NCA's operational focus. Peddie, 'Anti-terrorism legislation', p. 447.

28. NCA Annual Plan 2017/18, NCA, March 2017.
29. As can the SFO and the FCA.
30. The National Strategic Assessment of Serious and Organised Crime, NCA, 2018, p. 5.
31. Within the task force is an Operations Group which consists of representatives from the NCA, HMRC, City of London Police, Metropolitan Police Service, SFO, FCA, CIFAS, and staff from Barclays, BNP Paribas, Citigroup, Deutsche Bank, JPMorgan, HSBC, Lloyds, Metro Bank, Nationwide, Post Office, RBS, Santander and Standard Chartered.
32. Suspicious Activity Reports (SARS), NCA, Annual Report 2017.
33. NCA Annual Report and Accounts, NCA, 2016/17. Also see *National Risk Assessment of Money Laundering and Terrorist Financing 2017 (NRA 2017)*, HM Treasury and Home Office, p. 11.
34. NCA Annual Report and Accounts, NCA, 2016/17. Also see National Risk Assessment of Money Laundering and Terrorist Financing 2017 (NRA 2017), HM Treasury and Home Office, p. 11.
35. Statement by John Glenn, Commons HCWS1162, 10 Dec. 2018.
36. UK Mutual Evaluation Report op. cit., p. 6,
37. Ibid. p. 40.
38. Ibid. p. 53
39. Glenn, op.cit.
40. UK Mutual Evaluation Report, op. cit., p. 8.
41. *Suspicious Activity Reports 2017*, NCA, p. 40.
42. *National Risk Assessment of Money Laundering and Terrorist Financing 2015 (NRA 2015)*, HM Treasury and Home Office, p. 91.
43. Written Ministerial Statement, Operation of the UK's Counter-Terrorist Asset Freezing Regime, 1 Apr. 2017 to 30 June 2017. For a list of persons designated for terrorism and terrorist financing see https://www.gov.uk/government/publications/current-list-of-designated-persons-terrorism-and-terrorist-financing Last Accessed 29 July 2018. Under the Terrorism Act 2000, the Home Secretary may proscribe an organisation if he or she believes it is concerned in terrorism, and it is proportionate to do. See Home Office, Proscribed Terrorist Organisations. https://www.gov.uk/government/publications/proscribed-terror-groups-or-organisations—2 Last Accessed 29 July 2018.
44. Ibid. p. 103.
45. Criminal Finances Bill, Factsheet, Part 2: Terrorist Finance, https://www.gov.uk/government/publications/criminal-finances-bill-factsheets.
46. *NRA 2015*, p. 90.
47. Ibid.
48. UK Mutual Evaluation Report, op.cit., p. 94.
49. FCA Press Release, 31 Jan. 2017, https://www.fca.org.uk/news/press-releases/fca-fines-deutsche-bank-163-million-anti-money-laundering-controls-failure.

50. There were also about $3.8 billion in suspicious one-sided trades involved in the overall scheme.
51. FCA Final Notice to Deutsche Bank AG, 30 Jan. 2017, https://www.fca.org.uk/publication/final-notices/deutsche-bank-2017.pdf.
52. Office for Professional Body Anti-Money Laundering Supervision (OPBAS), FCA, 30 Apr. 2018.
53. *Anti-Money Laundering Annual Report 2016/17*, Financial Conduct Authority, p. 10.
54. Ibid., p. 10.
55. Ibid., p. 13.
56. Anti-Money Laundering Annual Report, op. cit, p. 32.
57. Financial crime analysis of firm's data, FCA, November 2018.
58. Ibid., p. 20.
59. Ibid., p. 4.
60. UK Mutual Evaluation Report, op. cit., p. 139.
61. 'Remarks in the Rose Garden by President George W. Bush', *Washington Post*, 24 Sept. 2001.
62. Currency and Foreign Transaction Reporting Act of 1970 as amended.
63. *US Anti-Money Laundering and Trade Sanctions Rules for Financial Institutions, Practical Law Finance*, Thomson Reuters, 2018.
64. In the United States there are mandatory and discretionary SAR filings with banks being protected from liability for any voluntary SAR filings they make.
65. *US Anti-Money Laundering and Trade Sanctions Rules.*
66. Other pieces of legislation of import here are the International Money Laundering Abatement and Anti-Terrorist Financing Act of 2001 (Title III of the USA PATRIOT Act, P.L. 107–56); Suppression of the Financing of Terrorism Convention Implementation Act of 2002 (Title II of P.L. 107–197); Intelligence Authorization Act for Fiscal Year 2004 (P.L. 108–177); Intelligence Reform and Terrorism Prevention Act of 2004 (P.L. 108–458); Combating Terrorism Financing Act of 2005 (Title IV of P.L. 109–177); and Implementing Recommendations of 9/11 Commission Act of 2007 (P.L. 110–53).
67. *United States, Mutual Evaluation Report*, FATF, Dec. 2016, p. 23.
68. These include the office of the Comptroller of the Currency, the Federal Reserve Board and the Federal Deposit Insurance Corporation.
69. The Department of Health and Human Services has responsibility for combating healthcare fraud and connected money laundering.
70. TFI also implements economic sanctions developed by Treasury's OFAC and the Department of State's Bureau of Counterterrorism.
71. These agencies report to FinCEN. They have produced a manual setting out their requirements. The agencies may issue cease and desist orders or choose more informal measures. See Chapter 4.

72. Also relevant here are the Organized Crime Drug Enforcement Task Force, the Internal Revenue Criminal Investigation, Immigration and Customs Enforcement, the El Dorado Task Force, the US Coast Guard, the US Secret Service and the US Postal Inspection Service.

73. Also relevant here are the Federal Deposit Insurance Corporation, the Office of the Comptroller of the Currency, the National Credit Union Administration and the State Banking Regulators. There are others with responsibility for AML and CFT compliance in the securities, futures and derivatives sectors and also for supervising casinos, non-profit organisations and other sectors.

74. Appendix G: 'Could America do more? An examination of U.S. efforts to stop the financing of terror', 9 Sept. 2015 in *Stopping Terror Finance*, US House of Representatives, 114 Congress First Session.

75. David S. Cohen, 4 March 2014, available at https://www.treasury.gov/press-center/press-releases/Pages/jl2308.aspx.

76. FATF, *US Mutual Evaluation Report*, p. 3.

77. Ibid.

78. Ibid., p. 6.

79. Ibid., p. 91.

80. Ibid., p. 4.

81. Ibid., pp. 53–4.

82. Ibid., p. 65.

83. Ibid., p. 87.

84. For the Foreign Corrupt Practices Act see https://www.justice.gov/criminal-fraud/statutes-regulations. For the Bribery Act see http://www.legislation.gov.uk/ukpga/2010/23/pdfs/ukpga_20100023_en.pdf. For an overview of the Bribery Act 2010, see *Bribery and Corruption UK Guide*, Ashurst, 12 Jan. 2017.

85. 95th Congress 1st Session, House of Representative Report No. 95–640, Unlawful Corporate Payments Act of 1977, 28 Sept. 1977, https://www.justice.gov/sites/default/files/criminal-fraud/legacy/2010/04/11/houseprt-95-640.pdf.

86. Jim Zarroli, 'Trump used to disparage an anti-bribery law: Will he enforce it now?', NPR, 8 Nov. 2017.

87. The International Corruption Unit of the NCA is responsible for investigating corruption, international bribery and money laundering including money laundering in the United Kingdom resulting from the corruption of overseas PEPs. Other relevant agencies include the Serious Fraud Office, the FCA and the NCA.

88. Under the Bribery Act an individual found guilty may be liable to imprisonment for up to ten years and/or an unlimited fine while a company found guilty will be subject to an unlimited fine. Under the FCPA an individual can be fined $250,000 per violation and may face five years' imprisonment. A company may be liable for $2 million per violation.

89. 'Differences between UK Bribery Act and the US Foreign Corrupt Practices Act', Norton Rose Fulbright, June 2011.

90. FinCEN also now have a beneficial ownership disclosure rule. There is no central registry but there is a requirement for disclosure so that financial institutions can perform due diligence. I am indebted to John Lawlor for pointing this out to me.

91. D. Sabbagh and H. Stewart, 'May faces cross-party push for public company registers in overseas territories', *The Guardian*, 27 April 2018.

92. The Crown Dependencies of the Isle of Man, Jersey and Guernsey will not be required to make their ownership registers publicly available but it is to be expected that they will come under pressure to voluntarily do so.

6. FINANCIAL SANCTIONS AND EXPORT CONTROL REGIMES

1. Javed Ali, 'Chemical weapons and the Iran-Iraq War: A case study in non-compliance', *The Non Proliferation Review*, Spring 2001, pp. 47–8.

2. Gareth Porter, 'When the Ayatollah said no to nukes', *Foreign Policy*, 16 Oct. 2014, p. 3.

3. Leonard S. Spector with Jacqueline R. Smith, *Nuclear Ambitions*, Westview Press, 1990, p. 211.

4. For a review of the Iranian nuclear programme, see Gary Sarmore (ed.), *The Iran Nuclear Deal: A Definitive Guide*, Harvard Kennedy School, Belfer Center for Science and International Affairs, 3 Aug. 2015.

5. See UN Chapter VII, Article 41, http://www.un.org/en/sections/un-charter/chapter-vii/.

6. For the list of sanctions committees, see United Nations Security Council Subsidiary Organs, https://www.un.org/sc/suborg/en/sanctions/information. Each committee is supported by a panel of experts.

7. https://www.un.org/sc/suborg/en/sanctions/un-sc-consolidated-list.

8. For a summary of each sanctions program or embargo see https://www.treasury.gov/resource-center/faqs/Sanctions/Pages/ques_index.aspx Last Accessed 5 Aug. 2018.

9. https://www.treasury.gov/resource-center/sanctions/Programs/Pages/Programs.aspx.

10. https://www.treasury.gov/resource-center/sanctions/Pages/regulations.aspx.

11. https://www.treasury.gov/resource-center/sanctions/Pages/licensing.aspx.

12. https://www.treasury.gov/resource-center/faqs/Sanctions/Pages/faq_general.aspx#basic.

13. Ibid.

14. For current penalty amounts see section V.B.2.a of Appendix A to OFAC's Economic Sanctions Enforcement Guidelines at 31 C.F.R. Part 501.

15. There is an exception for compliance advice.

16. See https://eeas.europa.eu/headquarters/headquarters-homepage/423/sanctions-policy_en.

17. See *EU Sanctions in Force*, http://ec.europa.eu/dgs/fpi/documents/Restrictive_measures-2017-08-04-clean_en.pdf; and *Consolidated List of Persons, Groups and Entities Subject to EU Financial Sanctions*, https://data.europa.eu/euodp/data/dataset/consolidated-list-of-persons-groups-and-entities-subject-to-eu-financial-sanctions.

18. EU law contains a defence that you 'did not know nor had reasonable cause to suspect' that you were dealing with a DP.

19. See Chapter 5.

20. New guidance on financial and trade sanctions for importers and exporters, OFSI, 11 May 2018. https://www.gov.uk/government/news/new-guidance-on-financial-and-trade-sanctions-for-importers-and-exporters Last Accessed 5 May 2018. The NCA investigates breaches of financial sanctions and HMRC enforces breaches of trade sanction. The Home Office implements any travel bans.

21. Annual Review, *OFSI*, Apr. 2017–Mar. 2018, p. 2.

22. The Policing an Crime Act 2017 has sought to speed up the implementation by the United Kingdom of implementation of United Nations listing so as to avoid asset flight.

23. https://www.gov.uk/government/publications/financial-sanctions-consolidated-list-of-targets Last Accessed 5 Aug. 2018. Note that the list does not include proscribed organisations under TA 2000 as not all are subject to sanctions.

24. Annual Review, OFSI, op. cit, p. 2.

25. https://www.gov.uk/government/publications/financial-sanctions-consolidated-list-of-targets/ukraine-list-of-persons-subject-to-restrictive-measures-in-view-of-russias-actions-destabilising-the-situation-in-ukraine 2018.

26. For details see Financial Sanctions Guidance, OFSI, Mar. 2018, pp. 17–21. https://assets.publishing.service.gov.uk/government/uploads/system/uploads/attachment_data/file/685308/financial_sanctions_guidance_march_2018_final.pdf Last Accessed 5 Aug. 2018. For guidance on reporting suspected breaches to OFSI see https://www.gov.uk/guidance/suspected-breach-of-financial-sanctions-what-to-do Last Accessed 5 Aug. 2018. Note that reporting requirements to OFSI are in addition to other reporting requirements including SARS.

27. This is set out in the Policing and Crime Act 2017, http://www.legislation.gov.uk/ukpga/2017/3/contents/enacted.

28. OFSI, *New Guidance*, p. 4.

29. Annual Review OFSI, op. cit., p, 3.

30. Ibid. p. 5. Also OFSI licenses cannot be issued retrospectively. There may be humanitarian and medical exemptions.

31. There are various professional tools for checking organisations and individuals against sanctions lists including those produced by Dow Jones and LexisNexis.

32. For an overview of US sanctions on Iran, see Kenneth Katzman, 'Iran sanctions', CRS, 29 June 2018.

33. https://www.treasury.gov/press-center/press-releases/Pages/hp644.aspx.

34. US Department of the Treasury, Press Center, Fact Sheet: New Sanctions on Iran, 21 Nov. 2011, https://www.treasury.gov/press-center/press-releases/Pages/tg 1367.aspx.

35. These regulations were originally introduced in 1987 and have been tightened over the years.

36. Also see the National Defense Authorization Act for Fiscal Year 2012.

37. Katzman, 'Iran sanctions', p. 28.

38. Ibid., p. 44.

39. https://eur-lex.europa.eu/legal-content/EN/TXT/?uri=CELEX%3A31996R 2271.

40. In the United Kingdom for example, already in 1992 the country had adopted the Protection of Trading Interest (US Cuban Assets Control Regulations) Order 1992 (SI 1992 No. 2449) which required that United Kingdom persons not comply with American prohibitions on trading with Cuba. In 1996 the United Kingdom amended this order in the UK Extraterritorial US Legislation (Sanctions against Cuba, Iran and Libya) (Protection of Trading Interests Order) Order 1996 so as to comply with the Blocking Regulation.

41. 'Austria charges bank after Cuban accounts cancelled', Reuters 27 April 2007.

42. Also relevant in the context of opposing extraterritorial action is German anti-boycotting legislation that arose in response to efforts by some Arab countries to boycott Israel. See section 7 of the German Foreign Trade Ordinance (Außenwirtschaftsverordnung). See 'Dealing with sanctions and anti-boycott measures under German and European law in financing transactions', Watson, Farley and Williams, Aug. 2016.

43. Remarks by Secretary of the Department of the Treasury, Jacob J. Lew, at the Washington Institute for Near East Policy, 29 April 2015, quoted in Katzman, 'Iran sanctions', p. 58.

44. Clifford Kraus, 'US decision could rattle oil markets', *International New York Times*, 11 May 2018.

45. For details of what arms control steps Iran was prepared to make in exchange for sanctions relief see The Iran Nuclear Deal: A Definitive Guide op. cit.

46. Statements on behalf of the EU, European Union External Action, 19 May 2016.

47. UNSCR 2231 also established a 'snapback' mechanism by which lifted sanctions could be re-imposed as a result of Iranian non-compliance with the JCPOA.

48. Thomas Erdbrink, 'Iran sanctions leave foreign firms bitter', *International New York Times*, 29 May 2018.

49. Countering America's Adversaries Through Sanctions Act" (Public Law 115–44) (CAATSA) https://www.treasury.gov/resource-center/sanctions/Programs/Documents/hr3364_pl115–44.pdf Last Accessed 5 Aug. 2018.

50. https://www.treasury.gov/resource-center/sanctions/Programs/Documents/jcpoa_winddown_faqs.pdf.

51. Joint statement on the Re-imposition of US Sanctions on Iran, 6. Aug. 2018, https://www.gov.uk/government/news/joint-statement-on-the-re-imposition-of-us-sanctions-on-iran.

52. Creation of proposed SPV to protect European Companies from US sanctions against Iran continues to be delayed, Baker McKenzie, Nov. 28, 2018.

53. http://www.legislation.gov.uk/ukpga/2018/13/contents/enacted.

54. There are due process requirements for designations, including notifying the designated parties and reviewing their status every three years.

55. As noted, sanctions may also be applied to support compliance with international human rights laws. Powers within the Criminal Finance Act 2017 permit authorities to seize any asset which may be linked to this type of conduct.

56. It can be noted that new OFSI-style licensing powers give OFSI latitude to grant broader sanctions exemptions than permitted under either United Nations or European Union regimes.

57. In addition, under SAMLA, sanctions-reporting obligations will be widened (together with the relevant criminal offence) to all natural and legal persons. Compliance obligations will thus similarly increase.

58. Neil MacFarquhar, 'Isolation of Russia is failing to shake Putin', *International New York Times*, 19 April 2018.

59. Common Military List of the European Union adopted by the Council on 6 March 2017 (equipment covered by Council Common Position 2008/944/CFSP defining common rules governing the control of exports of military technology and equipment), https://www.fdfa.be/sites/default/files/atoms/files/20180226_Common%20Military%20List%20of%20the%20EU.PDF.

60. https://eur-lex.europa.eu/legal-content/EN/TXT/?uri=CELEX%3A32009R0428.

61. EU Dual-Use List Annexes I & IV to the EU Regulation 428/2009, https://eur-lex.europa.eu/legal-content/EN/TXT/?uri=CELEX%3A32009R0428.

62. The full list is Category 0 Nuclear Materials, facilities and equipment; Category 1 Special material and related equipment; Category 2 Material processing; Category 3 Electronics; Category 4 Computers; Category 5 Telecommunications and Information Security; Category 6 Sensors and Lasers; Category 7 Navigation and Avionics; Category 8 Marine; and Category 9 Aerospace and Propulsion. Each category is divided into subcategories.

63. The relevant EU licence here is the EU General Export Authorisation license.

64. The UK Dual-Use List (Schedule 3 to UK Export Control Order 2008), https://www.gov.uk/guidance/controls-on-dual-use-goods.

65. The UK Military List (Schedule 2 to UK Export Control Order 2008), https://assets.publishing.service.gov.uk/government/uploads/system/uploads/attachment_data/file/685044/controllist20180305.pdf.

66. These lists all reflect the imprint of the Australia Group, Chemical Weapons Convention, Missile Technology Control Regime, the various lists of the Nuclear

Suppliers, and the various lists of the Wassenaar Arrangement. The UK Strategic Control Lists are drawn, inter alia, from the following: the Export Control Order 2008 (ECO 2008) SI 2008/3231, as amended; Schedule 2, last amended by SI 2018/165 [UK Military List], Articles 4A and 9 [UK Security and Human Rights List]; Schedule 3, last amended by SI 2017/85 [UK Dual-Use Control List]; Annexes II and III of Council Regulation (EC) No. 1236/2005, last amended by Regulation No. 775/2014 [EU Human Rights List]; the Export of Radioactive Sources (Control) Order 2006 S.I. 2006/1846, Schedule [UK Radioactive Sources List]; Annexes I and IV of Council Regulation (EC) No. 428/2009, last amended by Regulation (EU) No. 2268/2017 [EU Dual-Use List].

67. The UK Military List, for example, is divided between three categories (A, B and C) with control of Category A items being the most stringent. Also strictly controlled is where brokering touches upon embargoed areas.

68. https://www.gov.uk/guidance/european-union-general-export-authorisations.

69. https://www.gov.uk/government/collections/open-general-export-licences-ogels.

70. https://www.gov.uk/guidance/standard-individual-export-licences.

71. https://www.gov.uk/guidance/licence-types-faqs. Last Accessed 5 Aug. 2018. Note that there are also equivalent trade control licenses.

72. In the United Kingdom trade licence applications can be applied for online through ECJU's online exporting licence system, https://www.spire.trade.gov.uk/spire/fox/espire/LOGIN/login. This does not apply for financial sanctions.

73. In the United Kingdom this includes reasonable grounds to suspect WMD use.

74. See https://www.gov.uk/government/publications/guidance-on-the-supplementary-wmd-end-use-controls-including-provision-of-technical-assistance.

75. https://www.gov.uk/guidance/current-arms-embargoes-and-other-restrictions.

76. See Supplement 1 to EAR Part 774, https://www.bis.doc.gov/index.php/documents/regulation-docs/435-part-774-the-commerce-control-list/file.

77. If a company's item has a USML number, then the company must be registered with the State Department's Directorate of Defense Trade Controls (thus becoming ITAR-certified) and the item will require a licence.

78. When an item is located within the CCL it is provided with a classification number that describes that item and indicates the licensing requirements. The CCL is divided into ten categories (0: nuclear and miscellaneous; 1: materials, chemicals, etc.; 2: material processing; 3: electronics; 4: computers; 5: telecommunications and information security; 6: sensors and lasers; 7: navigation and avionics; 8: marine; and 9: aerospace and propulsion). Each category is then divided into functional groups (systems, equipment and components; test, inspection and production equipment; material, software and technology). Items which are not listed on the CCL are known as EAR 99 Items and may—subject to end-use considerations—as a general rule be exported anywhere without a licence, though US persons may not be able to export them to sanctions targets.

79. The Bureau of Industry and Security is part of the US Department of Commerce.

80. The White House, 'Fact sheet: Ukraine-related sanctions', 17 March 2014, https://obamawhitehouse.archives.gov/the-press-office/2014/03/17/fact-sheet-ukraine-related-sanctions.

81. Ukrainian Freedom Support Act of 2014 (P.L. 113–272) as amended by P.L. 115–44; the Support for the Sovereignty, Integrity, Democracy and Economic Stability of Ukraine Act of 2014 (P.L. 113–95); P.L. 115–44, Countering America's Adversaries through Sanctions Act, Title II. See https://www.congress.gov/113/plaws/publ95/PLAW-113publ95.pdf.

82. Including EAR 99 items.

83. Gardiner Harris, 'U.S. to issue new sanctions on Russia over Skripal's poisoning', *New York Times*, 8 Aug. 2018.

84. Even prior to the Ukrainian crisis, the United States was already sanctioning Russian individuals. In 2012 Congress had enacted the Sergei Magnitsky Rule of Law Accountability Act of 2012 (title IV, P.L. 112–208; 22 U.S.C. 5811 note) which required the President to identify the person(s) involved in the death of the lawyer Sergei Magnitsky. Over forty individuals are subject to Magnitsky sanctions.

85. Rebecca M. Nelson, 'US sanctions and Russia's economy', CRS, 17 Feb. 2017, p. 11.

86. See 'EU sanctions against Russia over Ukraine crisis', https://europa.eu/newsroom/highlights/special-coverage/eu-sanctions-against-russia-over-ukraine-crisis_en; and European Council, 'EU restrictive measures in response to the crisis in Ukraine', http://www.consilium.europa.eu/en/policies/sanctions/ukraine-crisis.

87. http://www.consilium.europa.eu/en/policies/sanctions/ukraine-crisis.

88. See https://www.gov.uk/government/publications/financial-sanctions-ukraine-sovereignty-and-territorial-integrity. Council Regulation (EU) No 269/2014 on restrictive measures against actions undermining or threatening the territorial integrity, sovereignty and independence of Ukraine. https://eur-lex.europa.eu/legal-content/EN/TXT/?uri=CELEX%3A32014R0269; Council Regulation (EU) No 833/2014 on restrictive measures in view of Russia's actions to destabilise the situation in Ukraine. https://eur-lex.europa.eu/legal-content/EN/TXT/?uri=CELEX:32014R0833. The most recent amendments to the Regulation can be found by searching on EUR-Lex https://eur-lex.europa.eu/homepage.html Last Accessed 5 Aug. 2018.

89. https://ec.europa.eu/growth/content/note-eu-business-operating-crimea-and-sevastopol_en.

90. See fn. 23. Also see The Ukraine (European Union Financial Sanctions) (No. 2) Regulations 2014. http://www.legislation.gov.uk/uksi/2014/693/contents/made; The Ukraine (European Union Financial Sanctions) (No. 3) Regulations 2014.

http://www.legislation.gov.uk/uksi/2014/2054/contents/made; Last Accessed 5 Aug. 2018. Also see the various updates.

91. The EU's political and economic relations with Ukraine are governed by the Association Agreement, which includes a free trade agreement. In August 2014 Russia proclaimed a retaliatory ban on the import of certain foodstuffs from the European Union and the United States though this too had no effect on Western willingness to continue with the mixture of sanctions and embargoes.

92. Nelson, 'US sanctions and Russia's economy', pp. 4–6.

93. Ibid., pp. 5–6.

94. Ambassador Nikki Haley, 'Remarks at a UN Security Council briefing on Ukraine', US Mission to the United Nations, 2 Feb. 2017, https://usun.state.gov/remarks/7668.

95. Thucydides, *The History of the Peloponnesian War*, ch. XVII.

96. Ross L. Denton, *EU Blocking Regulation, Sanctions and Export Controls Update*, Baker & McKenzie, 18 May 2018.

97. House of Lords, European Union Committee, 8th Report of Session 2017–19, Brexit: Sanctions Policy, 17 Dec. 2017, p. 34, para 130.

7. COUNTER-PROLIFERATION FINANCE

1. Mark Schapiro, 'The middleman', *Mother Jones*, May/June 2005; David S. Cloud, 'US says banned nuclear technology went to Pakistan and India', *New York Times*, 9 April 2005.

2. J.C. Zarate, *Treasury's War*, Public Affairs, 2013, pp. 427–8.

3. John Innes, 'Bin Laden admits he '"instigated"' US embassy attacks', *Scotsman*, 4 Jan. 1999. Also see K. McCloud and M. Osborne, 'WMD terrorism and bin Laden', Middlebury Institute of International Studies at Monterey, 14 July 2008.

4. Martin Navias and Rory Miller, 'Al Qaeda and weapons of mass destruction: What we know; how we know it', *World Defence Systems: The International Review of Defense Acquisition Issues*, Vol. 8, No. 1 (2005), pp. 137–9.

5. Michael Laufer, 'A.Q. Khan nuclear chronology', Carnegie Endowment for International Peace, 7 Sept. 2005.

6. The proliferation financing provisions to be found in the UNSCRs are reproduced in Annex C of the FATF 2018 CPF Report, See fn. 24 below.

7. *Proliferation Financing Report 2008*, FATF, 18 June 2008.

8. Ibid.

9. *Combating Proliferation Financing: A Status Report on Policy Development and Consultation*, FATF, 2010.

10. For case studies see *Proliferation Financing Report 2008*, pp. 24–42.

11. Ibid., p. 53.

12. UNSCR 2325 (2016) OP 7 calls upon states to take into account developments

in science and the evolving nature of proliferation risks and how these impact on their implementation of UNSCR 1540 (2004), http://unscr.com/en/resolutions/doc/2325.

13. *International Standards on Combating Money Laundering and the Financing of Terrorism & Proliferation: The FATF Recommendations*, Updated Feb. 2018, p. 11.

14. Immediate Outcome 11 requires that states without delay enact alterations to amendments in proliferation-related United Nations targeted financial sanctions. It also requires various public-private co-ordination measures.

15. Obligations under UNSCR 1540 (2004) do not form part of Recommendation 7, its Interpretative Notes nor Immediate Outcome 11. They do form part of FATF Recommendation 2.

16. *The Implementation of Financial Provisions of United Nations Security Council Resolutions to Counter the Proliferation of Weapons of Mass Destruction*, FATF, June 2013.

17. For an overview of North Korean sanctions, see for example Eleanor Albert, 'What to know about the sanctions on North Korea', CFR, 3 Jan. 2018; and also see Andrea Berger, 'House without foundations: The North Korea sanctions regime and its implementation', RUSI, 9 June 2017. For US sanctions see https://www.treasury.gov/resource-center/sanctions/Programs/pages/nkorea.aspx.

18. Report of the Panel of Experts established pursuant to resolution 1874 (2009), S/2018/171, 5 Mar. 2018. https://www.un.org/sc/suborg/en/sanctions/1718/panel_experts/reports Last Accessed 12 Aug. 2018.

19. Ibid., pp. 59–79.

20. See Sara Perlangeli, 'Flagging down North Korea on the high seas', RUSI, 29 March 2018. This brief article contains a useful discussion of some of the challenges faced in countering the maritime problem.

21. Jody Warrick, 'High seas shell game: How North Korean shipping ruse makes a mockery of sanctions', *Washington Post*, 3 March 2018.

22. Report of the Panel of Experts established pursuant to resolution 1874 (2009), S/2018/171, p. 4.

23. Ibid., p. 5.

24. *FATF Guidance on Counter Proliferation Financing: The Implementation of Financial Provisions of United Nations Security Council Resolutions to Counter Proliferation of Weapons of Mass Destruction*, FATF, Feb. 2018.

25. Ibid., p. 9.

26. Ibid., p. 13

27. Ibid., pp. 13–14.

28. For a full list of activity based prohibitions see Annex C to the 2018 CPF Report, Part (II)(c) to (i).

29. See Annex C, Part 11(k) for full 'vigilance' list.

30. See Annex C Part II(j) for the full list.

31. Ibid., p. 20.
32. Ibid., p. 25.
33. UK Mutual Evaluation Report, FATF, Dec. 2018, p. 107.
34. Ibid. p. 107.
35. 2018 CPF Report, op. cit., pp. 29–31
36. Ibid., pp. 32–4.
37. Aaron Arnold, 'Facing the myths surrounding proliferation financing', *Bulletin of the Atomic Scientists*, 11 April 2018.
38. Ibid.
39. Emil Dall, Andrea Berger and Tom Keatinge, 'Out of sight, out of mind: A review of efforts to counter proliferation finance', Whitehall Report 3–16, Finance, RUSI, June 2016, p. 16.
40. Ibid., p. 13. The authors also note that many governments are tardy in transposing UN sanctions designations into national legislation.
41. Ibid., p. 19.
42. Ibid., p. 26.
43. Emil Dall, Tom Keatinge and Andrea Berger, 'Countering proliferation finance: An introductory guide for financial institutions', RUSI, April 2017.
44. Also see Andrea Berger and Anagha Joshi, 'Countering proliferation finance: Implementation guide and model law for governments', RUSI, July 2017.
45. http://www.un.org/en/ga/search/view_doc.asp?symbol=S/RES/418(1977).
46. *Consultation on Proposed Changes to the FATF Standards, Compilation of Responses from the Financial Sector*, FATF, 2011, p. 167.
47. Ibid., pp. 66–7.

CONCLUSION

1. R.J. Heuer Jr., *The Psychology of Intelligence Analysis*, Center for the Study of Intelligence, CIA, 1999, p. 69. The author would also like to thank Patricia Asare.
2. David Aufhauser, General Counsel, Written Testimony before the House Financial Services Committee Subcommittee on Oversight and Investigations, 24 Sept. 2003, https://www.treasury.gov/press-center/press-releases/Pages/js758.aspx.
3. David Fein, 'How to beat the money launderers', *Financial Times*, 29 Nov. 2016.
4. Critics of the JCPOA of course argue that that agreement did not halt core Iranian proliferation behaviour.
5. Yaqoob Khan Bangash, 'Eating grass', *The Express Tribune*, 24 Jan. 2015.
6. Peter R. Neumann, 'Don't follow the money: The problem with the war on terrorism', *Foreign Affairs*, Vol. 96, No. 4, July/Aug. 2017.
7. M. Levitt and K. Bauer, D.C. Lindholm, C.B. Realuyo, J. Vittori and P.R. Neumann, 'Can bankers fight terrorism? What you get when you follow the money', *Foreign Affairs*, Nov./Dec. 2017.

8. Ibid.

9. Ibid.

10. The law firm did not say why it did not want to give evidence but it noted: 'We reject any suggestion based solely on the fact that we—like dozens of other international firms—operate in a particular market that our services may somehow involve the firm in corruption, state-related or otherwise.' Dan Sabbagh, 'Russian activity in City of London faces further scrutiny by MPs', *The Guardian*, 21 May 2018.

11. *Moscow's Gold: Russian Corruption in the UK*, House of Commons Foreign Affairs Committee, Eighth Report of Session 2017–19, p. 10.

12. Thomas Erdbrink, 'Europeans in Iran bitterly denounce the U.S. as "Caesar" amid "new sanctions"', *New York Times*, 26 May 2018.

13. R.S. Dunnett, C.B. Levy and A.P. Simoes, *The Hidden Costs of Operational Risk*, McKinsey & Co., Dec. 2005.

14. Jackie Calmes, 'Lew defends sanctions, but cautions on overuse', *New York Times*, 29 March 2016.

15. https://www.thefreelibrary.com/
Sir+Arthur+Harris+and+panacea+targets.-a0374335953.

16. J.C. Zarate, *Treasury's War*, Public Affairs, 2013, p. 428.

17. Carl von Clausewitz, *On War*, Wordsworth Classics, 1997, p. 5.

18. *Emerging Terrorist Financing Risks*, FATF, Oct. 2015, p. 44.

19. *Anti-Money Laundering and Counter-Terrorist Financing: Supervision Report 2015–17*, HM Treasury, 2018, p. 4.

20. *Moscow's Gold: Russian Corruption in the UK*, House of Commons Foreign Affairs Committee, p. 17.

21. See for example Martin Navias, 'Finance warfare and international terrorism', in *Superterrorism: Policy Responses*, edited by L. Freedman, Blackwell, 2002, pp. 57–79.

22. A. Ray, 'Dawn of a new era in AML technology', *Celent*, June 2018.

23. Ibid.

24. B. Patel, 'AI implementation in AML at HSBC sees a considerable reduction in compliance costs', *Finextra*, 16 Nov. 2017.

25. J. Fruth, 'Anti-money laundering controls failing to detect terrorists, cartels and sanctioned states', Reuters, 14 March 2018.

26. Fruth proposes adopting an actor-centric hybrid threat finance model for more finely attuned identification of emergent multi-dimensional money-laundering and terrorist-financing risks.

INDEX

INDEX